# WATERSHED HEALTH MONITORING

## Emerging Technologies

# WATERSHED HEALTH MONITORING

## Emerging Technologies

Chris Jones • R. Mark Palmer
Susan Motkaluk • Mike Walters

LEWIS PUBLISHERS

A CRC Press Company
Boca Raton London New York Washington, D.C.

**Cover photo:** Nottawasaga River near Wasaga Beach, Ontario, Canada. (Photo taken by Chris Jones and provided courtesy of the Nottawasaga Valley Conservation Authority, Angus, Ontario, Canada.)

**Library of Congress Cataloging-in-Publication Data**

Watershed health monitoring : emerging technologies / Chris Jones ... [et al.].
p. cm.
Includes bibliographical references (p. ).
ISBN 1-56676-967-1 (alk. paper)
1. Watershed management. 2. Environmental monitoring. I. Jones, Chris, 1969-

TC409 .W3696 2002
363.739'463--dc21 2002016118



Direct all inquiries to CRC Press LLC, 2000 N.W. Corporate Blvd., Boca Raton, Florida 33431.



**Visit the CRC Press Web site at www.crcpress.com**


Lewis Publishers is an imprint of CRC Press LLC


International Standard Book Number 1-56676-967-1
Library of Congress Card Number 2002016118
Printed in the United States of America 1 2 3 4 5 6 7 8 9 0
Printed on acid-free paper

# Dedication

*We dedicate this book to our colleagues and to our soon-to-be colleagues, the students who are looking to future careers in the fields of resource management, engineering, and life sciences. We also dedicate this book to our long-suffering spouses, friends, and relatives who had the strength to put up with us while we developed the manuscript.*

**—Chris Jones, R. Mark Palmer, and Mike Walters**

*I would like to acknowledge the incredible support, inspiration, encouragement, and motivation that I have received from Reginald Ventura, who in my life has been and continues to be my source of never-ending strength. I would also like to recognize the invaluable advice and wisdom that Rick Watts has so generously given to me over the years. Thank you both for helping me reach for my dreams. I dedicate this book to every woman who has had the courage to follow her dreams, to every woman who has walked before me, and to every woman who will follow.*

**—Susan Motkaluk**

# Foreword

The last decade has witnessed the emergence of the watershed as the logical unit for managing surface water and indeed, to support the management of groundwater. This was not always the case. For the last half century, since the concept of Ontario, Canada's watershed-based conservation authorities was realized, these agencies have sometimes labored under government ignorance and marginal interest. During this time, most decision makers have considered the watershed to be an irregular area of jurisdiction–one that does not fit into any heretofore human-developed administrative boundaries. Watershed organizations were considered special-purpose bodies, and watershed management was simply another planning and management process that could easily be accommodated within local or senior government planning systems.

Times change! In recent history, the public and its elected representatives have been constantly reminded by a continuous stream of unfortunate events that mankind has a significant capacity to despoil the water resource. Tragic events in Saskatchewan and Ontario, Canada have served to forcefully inform, if not shock, public officials into the realization that the systems designed to manage and protect water may very well be broken or, at best, not operating effectively. The reality in some locations may be much worse; they may not exist.

The watershed has come into its own. It finally has been discovered by the science community, public officials at all levels of government, and even the public. This discovery, although long overdue, will place significant pressure on watershed planners and managers to provide leadership, innovation, and a capacity for adaptation within the art and science of watershed management. Although the watershed concept has been discovered, it is essential to translate this interest into rational, logical processes that capture the interest of decision makers and (most important) the public if this movement is to result in sustained public good.

This book is a component of that required leadership and innovation. If watershed management is to be effective, it must be practiced. This book is written to inform those who seek to use the watershed approach to management. But it is not all about science, for there are many components to watershed management that are not strictly science based. Organization and structure, funding, public involvement, and decision making that transcends political boundaries are essential if effective watershed monitoring and management are to be realized. Watersheds are exceedingly intricate and interdependent complexes of land, water, plants, and animals. Systems that monitor and manage these complexes must reflect this complexity. This text gives us new tools for understanding such complexity and helps make the processes of watershed monitoring and management more ordered and understandable. Successful implementation, the ultimate goal, will be improved.

**James S. Anderson**
*Former General Manager, Conservation Ontario*
*Policy Advisor, Ontario, Ducks Unlimited Canada*

# *Acknowledgments*

We wish to acknowledge the contributions that the individuals and groups listed below made to this book. Their contributions took many forms. Some provided data for figures and tables. Others participated in case study projects. Still others provided ideas and editorial suggestions.

Jim Anderson, Ducks Unlimited Canada; Dr. Dave Barton, Dr. Mike Stone, Karen Trevors, and Dr. Colin Mayfield, University of Waterloo; Gus Rungis, Grand River Conservation Authority; Dr. Lyle Friesen, Canadian Wildlife Service; Ted Chesky, landbird consultant; Gary Bezruki, Denise Novotny, Karen Moyer, Greg Romanick, Sunda Siva, and Len Fay, City of Waterloo, Ontario, Canada; Brian Trushinski, City of Kitchener; Nicholas You, United Nations Centre for Human Settlements (Habitat); and Michael J. Rich, National Research Council of Canada.

Leanne Gelsthorpe, Centre for Research in Earth and Space Technology; Dan McGillivray, Centre for Research in Earth and Space Technology; Ray Blackport, Blackport Hydrogeology Inc.; Dwight Boyd, Grand River Conservation Authority; Hugh Simpson, Ontario Ministry of Agriculture, Food and Rural Affairs; Carol Wiebe, MacNaughton Hermson Britton Clarkson Planning Limited; Steve Holysh, Regional Municipality of Halton; Steven Knipping, MTE Consultants Inc.; and Mitch Tulloch, Vincent Massey Collegiate.

Nottawasaga Valley Conservation Authority, Angus, Ontario, Canada; 4DM Inc., Toronto, Ontario, Canada; the AEMOT Groundwater Management Area Steering Committee; members of the Stream Restoration Monitoring Framework review team; City of Tianjin Environmental Protection Bureau, China; City of Waterloo, Ontario, Canada; Environment Canada; Ganaraska Region Conservation Authority, Port Hope, Ontario, Canada; Greenland International Consulting Inc., Toronto, Ontario, Canada; Ji County Government, China; Kije Sipi Ltd, Ottawa, Ontario, Canada; Lake Simcoe Region Conservation Authority; Mississippi Valley Conservation Authority; Municipality of Clarington; Optech Inc.; Space Imaging Corporation; Sustainable Development & Monitoring Inc., Waterloo, Ontario, Canada; Town of Collingwood, Ontario, Canada; Ontario Ministry of Natural Resources; Stantec Consulting; Planning and Engineering Initiatives; Kitchener-Waterloo Homebuilders Association; Region of Waterloo, Ontario, Canada; Township of Uxbridge, Ontario, Canada; Dr. K.H. Nicholls, Lake Simcoe Environmental

Management Strategy Partnership; and Louise Morris, the Federation of Canadian Municipalities.

We wish to specially acknowledge the Nottawasaga Valley Conservation Authority for providing a facility and the required hardware to print the manuscript.

**Chris Jones**
**R. Mark Palmer**
**Susan Motkaluk**
**Mike Walters**

# About the Authors

**Chris Jones, B.Sc.,** began his career with the Nottawasaga Valley Conservation Authority in 1993. His primary responsibilities at the Conservation Authority are stream bioassessment and information management. He also provides consulting services in the areas of stream health monitoring, environmental project design, and related fields. Mr. Jones was educated at the University of Guelph, Ontario, Canada and is a member of the North American Benthological Society. He has published numerous works as "gray literature," coauthored several papers for research journals, and frequently contributes to an outdoor adventure magazine that is published in Ontario.

**R. Mark Palmer, B.Sc. (Eng.), P.Eng.,** has completed an impressive volume of assignments in 15 years of professional experience in hydrology, hydrogeology, hydraulics, and integrated watershed planning and management. He is a partner of Greenland International Consulting Inc., an environmental engineering company based in Toronto, Canada that provides innovative and sustainable solutions geared to the demands of five primary environmental sectors—water resources, environmental management, research and development, environmental monitoring, and information systems. A professional engineer and designated consulting engineer in Canada with a fisheries biology background, Mr. Palmer is affiliated with many professional associations around the globe. He has coauthored more than 20 publications that were published or presented at international conferences. Mr. Palmer has been recognized as an expert on water management and environmental planning and monitoring issues. He is married to Jane, has two sons, Andrew and Eric, and lives in Collingwood, Ontario, Canada.

**Susan Motkaluk, B.A.Sc., P.Eng.,** has an honors bachelor of applied science in civil engineering from the University of Waterloo, Ontario, Canada and is the president of Sustainable Development & Monitoring Incorporated (SDM). She has developed and implemented an award-winning, cost-effective partnering model for managing watersheds that has received worldwide attention and that is currently featured on the United Nations database as a Global Good Practice. As a result of her experiences with watershed management and the attention received from the United Nations, SDM created the Closed-Loop Model in an effort to educate watershed communities on maximizing watershed sustainability through monitoring, maintenance, and mitigation. Ms. Motkaluk's work incorporates elements of social, economic, political, and environmental theory. She is actively involved in the development of environmental policy, curriculum for students, community-based monitoring, and several innovative watershed research projects involving multi-stakeholder partnerships. She is continuing to bring her unique partnering model to local and global communities. Ms. Motkaluk has successfully built long-term partnerships with the local development industry, citizen groups, public agencies, and municipalities for collaborative work toward sustainable management of watersheds.

**Mike Walters, B.S.,** has worked for the Lake Simcoe Region Conservation Authority since 1984 and has been Manager of Environmental Services for the last 8 years. During his career, he has been involved in developing and implementing Conservation Authority programs and policies to protect and improve the health and quality of Lake Simcoe and its ecosystem. Mr. Walters has an honors degree in resource science from the University of Western Ontario and has extensive experience in water resource management, specifically the development of pollution control strategies and integrated watershed plans. He has coauthored a number of research and journal publications and is the chair of the Lake Simcoe Environmental Management Strategy Technical Committee. Mr. Walters recently returned from China, where he applied his expertise to assist the City of Tianjin in developing a nonpoint pollution control program. He is married to Kelly and resides within the Lake Simcoe basin in Bradford, Ontario, Canada.

# Contents

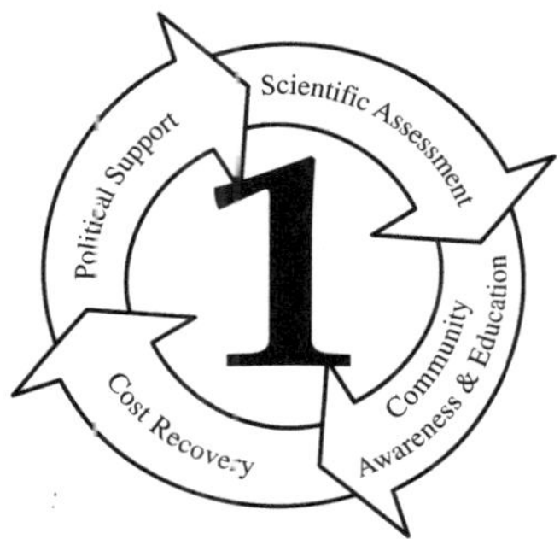


# Introduction

*Scientists who claim their work will solve global hunger, pollution, or overpopulation do not understand the social, economic, religious, and political roots of the problems that preclude scientific solution.... scientists have to get rid of the pernicious myth about the potential of their work to solve all our problems.*

**—Dr. David Suzuki[1]**

## 1.1 Problem statement

Our 42 years of collective experience working with resource management agencies and private consulting firms have revealed that successful monitoring programs are built on a foundation of four crucial components: political support; sound scientific assessment techniques; community education and involvement; and a long-term cost recovery framework. Too often, monitoring programs fail because they do not recognize the importance of each component, or the linkages between them. Even within each of these components, practitioners often run into trouble.

Within the realm of scientific assessment, for example, we have seen practitioners place much emphasis on surveillance (to answer questions about the status of watershed resources and how they change over time and space) while neglecting performance evaluation (which examines whether implemented management strategies and activities are resulting in a shift toward or away from stated goals). Generally reflective of their backgrounds in the life sciences, we have also seen many practitioners ignore the suite of human or socioeconomic aspects of watershed health (see Figure 1.1) by focusing assessments exclusively on environmental attributes. A monitoring program with all of the vital components still fails if it lacks a good feedback loop to managers, who should be using monitoring information for policy and program development as part of an iterative, adaptive watershed management process (Figure 1.2).

Concerns about effective watershed management and monitoring have been raised in the literature and around the world. At the International Workshop on River Basin Management (held in The Hague, the Netherlands,

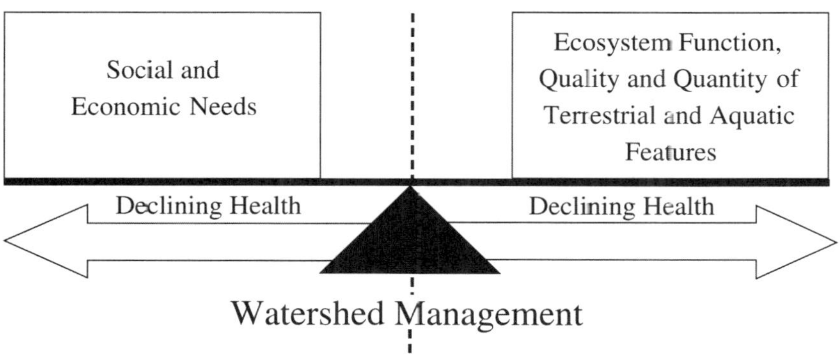


*Figure 1.1* Definition of watershed health. Watershed health is a state in which resource management activities sustain human needs and uses of the watershed while ensuring that ecological function is maintained. The status of health is illustrated here, with the watershed management fulcrum in a central position. Declining health, a state in which management favors either human uses or environmental issues, is illustrated here as a displacement of the watershed management fulcrum away from the central, or balanced, condition.

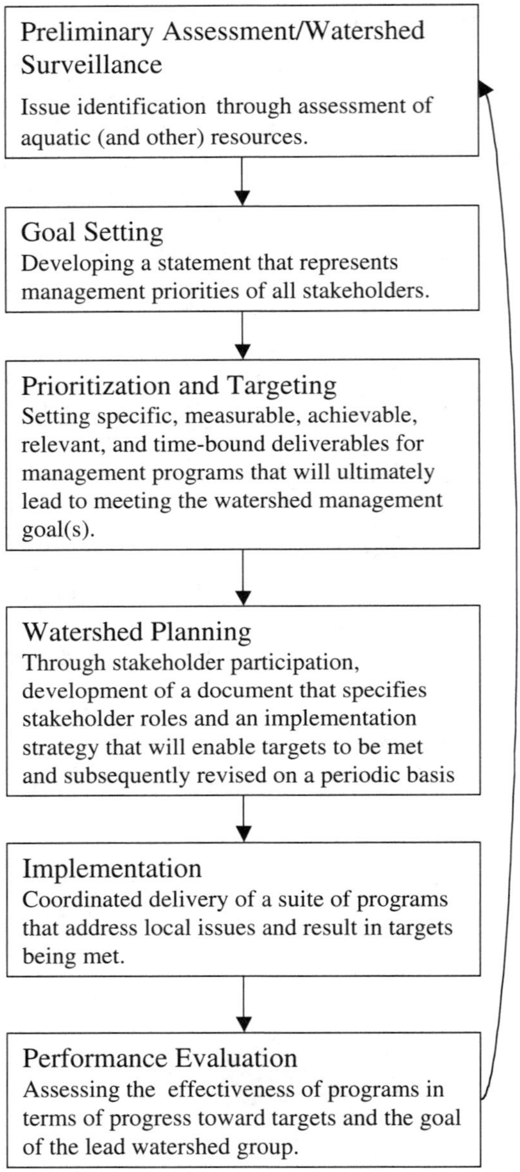


***Figure 1.2*** Watershed management—a cyclic process in which feedback from monitoring programs permits adaptation.

October 27–29, 1999), for example, delegates called for river basin management built on a foundation of effective leadership and political commitment; a bottom-up approach with local empowerment and effective public and stakeholder participation in decision making; acquisition and reporting of relevant data; and a sustainable funding model.[2] Similarly, the U.S. Intergovernmental Task Force on Monitoring Water Quality—which was formed in 1991 to evaluate the nation's monitoring programs—recommended improved reporting, performance evaluation, standardization of indicators, and assessment methodologies.[3]

With the population of the world more than 6 billion and projected to reach nearly 9 billion by 2050,[4] maintaining watershed health will be increasingly difficult. A watershed health monitoring model that helps practitioners to avoid common pitfalls is in order.

## 1.2 Background

All sectors of human society and nonhuman biota benefit equally from healthy watersheds. For this reason, people have been trying to assess spatial and temporal trends in watershed health for many years. Monitoring the health of watersheds is a necessity because so many of the activities that societies do as they grow and evolve conflict with each other and with the integrity of the natural environment. As our communities grow, for example, we pave watersheds and erect impervious structures that reduce infiltration, thereby short-circuiting the hydrologic cycle. We create impoundments on rivers, thereby changing trophic relationships, creating migration barriers for aquatic species, and altering thermal regimes. To provide food for our growing population, we clear land for agriculture. We harvest trees for lumber; pollute soils, water, and air with toxins and other contaminants; and transplant species.

Authorities throughout the world are focusing resource management activities and policies on watersheds. Watersheds represent natural boundaries and are linked with the geological history of the world, making them logical management units. Water is a fundamental resource for living things, so watersheds also represent habitat units. The hydrologic cycle of a watershed lends itself well to adaptive management because activities or imposed changes (e.g., hydrologic or habitat alterations) in one area of the catchment affect and can be detected at other areas.

In his foreword to the book *Watershed Management Practice, Policies and Coordination,* Eugene Odum[5] suggested that the move toward watershed management largely stemmed from our collective desire to move toward a more holistic, *ecosystem approach* to management. The ecosystem approach, he asserted, represents a much different approach for managers who conventionally treated resources (for example, soils, forests, and wildlife) in isolation and for short-term gains that fit well within our political systems.[5] The ecosystem is the lowest complete unit that provides all requisites for life and is hence the logical unit on which to focus theory and practice. The problem is that in populated areas, humans tend to fragment ecosystems to the point that management must be focused on the next intact level in the hierarchy of environmental organization, the landscape. Odum[5] therefore argued that, with defined boundaries and quantifiable inputs and outputs, the watershed is an ideal management unit within our fragmented ecosystems.

## 1.3 Objective and format

This book is the first in a series of pragmatic texts on watershed management. It is composed of two parts. The first part is a set of chapters (including this one) that define watershed health, describe the linkages between watershed management and monitoring, and detail a closed-loop model for successful watershed health monitoring. The second part is a collection of case studies that illustrate how our watershed health monitoring model has been successfully implemented.

Although we recognize the multidisciplinary nature of our watershed health definition (Figure 1.1), the focus of this book is on monitoring *aquatic* resources. We argue that because of the linkages among land, water, and air, the status of aquatic resources can be considered representative of overall watershed and ecosystem health. Other land- and air-based management and monitoring issues, as well as those dealing with the human side of the watershed health definition, will be dealt with in other texts in the series.

After reading this book, we hope that you will heed our advice by devoting appropriate time and resources to developing political, educational, and financial linkages as you dive into the science of watershed monitoring.

## 1.4 Definition of watershed health

We define watershed health as a measure of how well resource management activities are able to balance anthropogenic needs and ecological function and integrity within watersheds. As alluded to above, this balance is a difficult one to achieve because many human needs and uses of the environment are themselves ecological stressors.

Theoretically, the idea of sustainability is implicit in our definition because if the often-conflicting human and ecological needs are kept in balance, there is no net loss over time in the ability to maintain either one. The figure is intended to convey the notion that shifting management activities (the fulcrum) in favor of the social and economic needs (to the left) tends to result in a decline in the integrity of natural features. Conversely, too much of a focus on the natural features, and socioeconomic impacts will be felt. Either situation represents reduced watershed health.

Our watershed health model is in agreement with recommendations from the International Workshop on River Basin Management,[2] which stressed as a key message the following:

> The aim of sustainable river basin management is to ensure the sustained multi-functional use of the basin. Basic water needs of peoples and ecosystems should

be fulfilled first. Essential ecological and physical processes should be protected. Moreover, the effects on the receiving water bodies (seas, lakes, deltas, coastal zones) should be paid full attention.

## 1.5 The foundation of watershed management: A lead group and a plan

Critical to the management of watershed health is one *lead entity.* This group or agency must "be accountable to the public and all stakeholders, have sufficient financial and legislative means and work on the basis of a plan."[2] We should clarify that, although involvement of political leaders in the watershed management process is critical (see Chapter 2), the lead management entity does not necessarily need to be a governmental body. Particularly in the United States, there are many examples of successful watershed management in which the coordinating groups are collections of citizens or partnerships between government and stakeholder representatives.[6–10] The New Jersey Pinelands and Chesapeake Bay programs illustrate the importance of a lead agency that provides an open forum for the discussion of issues; promotes genuine partnerships between stakeholders; clearly expresses goals that address local needs and identified issues; and incorporates good research (monitoring) into the management framework.[11]

The lead agency must set goals and targets on the basis of consensus. In a multi-stakeholder process, balance between human and ecosystem needs cannot be achieved through balloting but must be achieved over time through a series of compromises. A process driven by popular vote is constantly threatened by the interests of a dominant stakeholder or a changing political climate. Both the leaders of watershed agencies and their management process must earn the trust of watershed stakeholders.[11] This trust empowers stakeholders to partner together and to develop and work toward common goals. Candidates for leadership must be found who possess the necessary skill set for the job. Elected officials are not necessarily the best people to lead watershed management efforts!

Many issues at the watershed management level are complex and interwoven. It is vital to avoid overlapping or conflicting mandates and gaps in authority or responsibility.[11] Hence, a job for the lead watershed group is defining roles and delegating responsibilities toward efficient implementation of the watershed plan. Enough flexibility must be built into the management framework to allow partner responsibilities to evolve. Just as management itself is an adaptive process, so too is stakeholder involvement. The lead watershed group must deliver a clear message to stakeholders early on regarding what an ecosystem is and what relationships exist between it and human activities and desires. Through integration and synthesis of stakeholder views, a clear vision must be presented by watershed leaders to achieve consensus and active cooperation between partners. The ability to break down problems and issues is central to this process. Large-scale, interrelated problems may seem insurmountable for participating stakeholders, but when broken down into components, they may be more easily understood and reconciled.[11]

Both the partners and the lead agency need persistence.[11] By its very nature, watershed management is a long-term process. A constant dilemma that managers face is that the long-term horizon of a good planning document generally meshes poorly with the short-term mentality of elected officials and government programs. The watershed agency needs to overcome this problem and avoid the common strategy of crisis management, whereby resources are allocated toward resolution of one crisis after another, according to the political climate of the day.[11] The performance evaluation aspect of watershed health monitoring programs grapples with this problem constantly. The challenge here is to report results even though they are exceedingly difficult to measure during the time frame over which people expect to see them. One way to sidestep this problem is to measure achievements via milestones that are linked with incremental progress toward goals.[11] On the other hand, it does not hurt to convey a sense of urgency to facilitate implementation. Indeed, most environmental legislation was formulated to address a crisis or conflict over resource allocation. Action is more likely if proactive management is at least partially sold as crisis management or avoidance.

During goal- and target-setting exercises, the lead agency needs to pay special attention to the role of science and policy. Although each of these influences needs to be taken into account in the management process, it is critical that they are not confused.

Once a lead watershed body is organized, the next critical step in watershed management is to develop a plan.

### 1.5.1 The watershed plan

The watershed plan is the foundation of watershed management. It assesses the current situation (including the identification of conflicts and priorities), formulates visions, sets goals and targets, and thus guides all aspects of management. It facilitates interaction and discussion between managers and stakeholders, offers a common point of reference, and thus provides coordination.[2] It is the starting point for generating public participation and—particularly relevant to watershed health monitoring—provides a framework for generating political support and funding. As an open process in which watershed stakeholders are encouraged to participate, the plan should lend credibility to, improve public acceptance of, and ultimately boost participation in management activities, including monitoring.[2]

A good watershed plan does not come together overnight. The amount of time required to develop the plan generally depends on several factors, including the size of the basin; the complexity of natural features and land use patterns within it; and the number of stakeholders and how opposed their views are. Partner consultation is vital to developing a watershed plan that reflects human needs and ecological function within the watershed. It also fosters a sense of ownership of the plan and facilitates "buy-in" from all local stakeholders. The key components of a watershed plan (see Table 1.1) include a background section that outlines watershed resources and issues; a description of the planning process; goal statements; watershed health targets (that address issues on either side of the watershed health equation); management actions; an implementation schedule and strategy; a long-term funding model that accounts for all management actions, including monitoring; a strategy for performance evaluation (which, as Chapter 3 describes, is one of two major components of the watershed health monitoring program); and a schedule for reviewing the plan.

***Table 1.1*** Principal Components of a Watershed Plan

| Section | Key content |
|---|---|
| Background | Overview of issues and resources on both sides of the watershed health equation: human needs and ecosystem function; inclusion of pertinent assessment data to establish benchmark watershed health status at time of writing |
| Watershed planning process | Details on the methodology for developing the plan and a matrix illustrating the roles and responsibilities of partners |
| Goal statement | Example: "The goal of the watershed management group is to ensure that cultural and socioeconomic functions of the watershed are maintained in balance with ecological integrity" |
| Watershed health targets | Specific, measurable, feasible, relevant, and time-bound targets that are reflective of local issues and the planning horizon of the document |
| Recommended management actions | A tabular or matrix format indicating key issues and recommended strategies or actions at a scale appropriate for implementation; in other words, management actions are often developed at the subwatershed scale |
| Implementation schedule and strategy | References for pertinent legislation and specification of the lead partner for each management action |
| Funding model | A detailed, long-term funding strategy for implementation of recommended management actions; performance evaluation and surveillance monitoring activities; and watershed plan revision |
| Performance evaluation | Methods for evaluation and reporting and a flowchart illustrating how performance information will be used to permit adaptive management |
| Schedule for watershed plan revision | Specification of the length of time between reviews of the *living document,* for instance, 5–10 years |

### 1.5.2 The watershed management model

In the U.S. Environmental Protection Agency's Watershed Approach Framework,[12] several guiding principles are highlighted as tenets for good watershed management. These include stakeholder involvement in management decisions; a geographic focus that is consistent with hydrologic boundaries (ground and surface watersheds); and an iterative (adaptive) decision process that is based on "strong science and data."

Golden[11] eloquently described why successful watershed management programs must be cyclic and adaptive, as illustrated in Figure 1.2:

> It is quixotic to believe that we will discover simple, elegant, and universal answers to resource management questions. There are two reasons for this. First, complexity, variability, and diversity are hallmarks of the natural environment. Second, at the same time, a wide spectrum of human interests extends across and intertwines within this far-from-simple environment. Therefore, management will be a perpetual and open-ended process of discovery and compromise, of substantial successes and backsliding. These are the realities of public resource management.

Monitoring is entrenched in watershed management because it begins and completes each cycle. The overall management process is generally started by some preliminary study to identify issues, collect historical data, conduct health assessments, and identify data gaps. Once targets are set, performance evaluation becomes important to determine whether implemented activities are successful. For example, if protection of an endangered species is a target, monitoring must track abundance and habitat of the species in question to document success or shortcomings of rehabilitation efforts. General surveillance (stream bioassessments, development statistics, etc.) over subsequent cycles generates trend data that are vital to ongoing prioritization of issues in the watershed. These activities answer questions such as the following:

- How does stream health change from year to year?
- How does the fishery resource change in a downstream direction through a given subwatershed?

Targets and milestones are usually modified over time using surveillance and performance information. In evaluating program successes, performance evaluation provides a feedback loop to managers and is the bridge to the next, adapted management cycle.

## 1.6 The Closed-Loop Watershed Health Monitoring Model

As our definition of watershed health (Figure 1.1) suggests, watershed health monitoring is a process that evaluates the state of balance between human uses of natural resources and ecological integrity within watersheds. We indicated in previous sections that watershed management, the watershed plan, and monitoring are inseparably linked. Figure 1.2 shows that monitoring (surveillance) data initiate the watershed management process and that performance evaluation provides a capacity for adaptation via a feedback loop to managers.[13] Watershed health monitoring is a complicated, interdisciplinary process that relies on four interrelated components, as shown in our Closed-Loop Model (Figure 1.3). The key components are community education and involvement; a direct link to and involvement of political decision makers; a sound scientific assessment component to quantify both sides of the model (human needs and ecological function); and a long-term cost recovery mechanism.[14]

Our rationale for including each of the four components in the Closed-Loop Model is summarized here and discussed in greater detail in the subsequent chapters. To start with, political support (Chapter 2) and involvement in monitoring activities is vital because it assists in the delivery of assessment information to policy and decision makers. It is also critical to funding because in most parts of the world, a significant component of the cost recovery framework for watershed management comes from government sources. Good scientific assessment techniques (Chapter 3) provide the basis for good decision making. They enable identification and prioritization

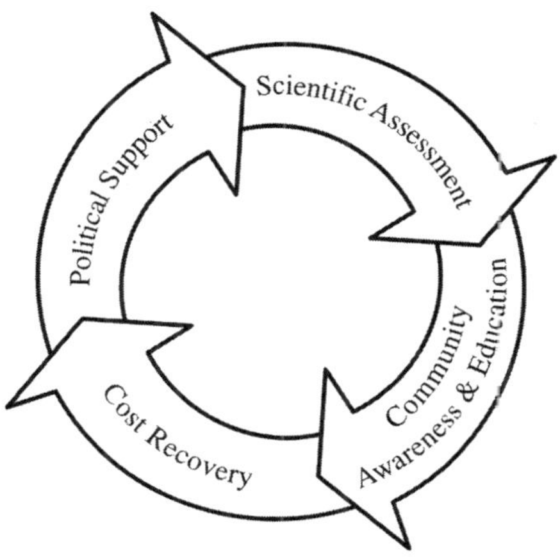


*Figure 1.3* The Closed-Loop Watershed Health Monitoring Model. The model illustrates that successful monitoring programs depend on four interrelated components: political support, scientific assessment, community awareness, and a mechanism for long-term cost recovery. (Adapted from models developed by Sustainable Development & Monitoring Inc., Waterloo, Ontario, Canada.)

of management issues and provide a means of evaluating performance. Just as "local empowerment and effective public and stakeholder participation in decision-making" were identified at The Hague[2] as critical attributes of the watershed management framework, community involvement, education, and partnering (Chapter 4) help to develop a bottom-up management process that reacts quickly to the needs of knowledgeable watershed stakeholders. Education and participation help to maintain momentum across all aspects of the Closed-Loop Model. A cost recovery model (Chapter 5) that is sustainable over the long term is necessary to support adaptive watershed management cycles that continue indefinitely.

# Political linkages and support

*River basin management is often characterized by parochial interests and intractable problems. To achieve progress, leadership and political commitment are essential.*

**—International Conference on River Basin Management[2]**

## 2.1 Rationale

A lead watershed group is required to coordinate all aspects of watershed management, including monitoring, and it is absolutely vital that this group or agency be intertwined with the political system of the region.[2] Likewise, each of the programs that form the overall management strategy within a given watershed must have political support and involvement to achieve success. In this chapter, we explore the reasons that we believe political linkages to be one of the four key components of successful watershed health monitoring programs (Figure 1.3) and discuss some methods of securing this support.

As we indicated in the previous chapter, the components of the Closed-Loop Model are interconnected. Because government sources (most commonly, local or regional administrations) are often one of the largest sources of funding for monitoring and other watershed management activities, it stands to reason that political support and involvement are critical to developing funding relationships that enable long-term delivery of monitoring (and other) programs.

We made the claim in Chapter 1 that lead watershed management groups must act on the basis of a plan, a document that contains targets and strategies related to issues exposed by monitoring. The implementation of watershed plans into local policy and legislative frameworks, however, is frequently problematic. Indeed, without strong political support and involvement, most management programs will fail, and there will be no need for performance evaluation or any opportunity to address issues raised by surveillance.

Some day-to-day aspects of monitoring programs are also dependent on links to local government decision makers. Surveillance indicators relating to the social and cultural side of the watershed health definition (Figure 1.1)—instantaneous, trend-over-time, and projection estimates related to population, employment, public health, and economic stability, for example—are typically collected by local governments. To minimize duplication, watershed managers need political involvement and support to facilitate sharing of data. To illustrate another example of this interdependence, consider that most watershed management plans contain targets for protection and rehabilitation of key aquatic and terrestrial habitats. The usual means of attaining such targets is through development setbacks and implementation of best management practices. It follows, then, that a critical precursor to good performance evaluation is access to local government data on where development has occurred, over what time frame, and on what controls were in place. Because government bodies and politicians also rely on performance evaluation to assure their constituents that local policies and programs are not proceeding at the expense of the environment, there are obvious political incentives associated with participation in monitoring efforts.

In addition to the core functions of surveillance and performance evaluation, there are a number of other management activities that rely on linkages between monitoring and political involvement and support. Monitoring data, for example, are required to assess surface and groundwater resources and to calibrate and verify water balance and water budget models used in developing surface or groundwater allocation frameworks. In this example, political support enables the decision framework to be incorporated into the local governance process. Similar linkages occur within the realms of enforcement and rehabilitation. Frequently, lead watershed management groups will not have the legislative backing to address all issues pertinent in their watershed and so will have to look at political partnerships with other agencies to round out their enforcement and compliance programs. Pollutant-trading programs[15] commonly embody such political linkages to do the following:

- Support early surveillance activities that expose excessive phosphorus loading as a local issue and provide a basis for establishing abatement targets
- Enable interagency partnerships, most often among government, the lead watershed agency, and private-sector business (agriculture, manufacturing, etc.)
- Maintain consistent regulation and enforcement to ensure program delivery in a fair and consistent manner across the watershed
- Facilitate adaptation by allowing past lessons to influence future decision making

The process of developing best management practices demonstrates linkages to other watershed management, monitoring, and politics. In this case, monitoring data characterize a common problem in the watershed, for example, contaminated agricultural runoff. Monitoring is also important to the development of cost-effective solutions such as containment facilities for manure and manure stack runoff; natural buffer areas between the sources of contamination and receiving water bodies; or treatment facilities such as lagoons, constructed wetlands, and grassed filter strips. Political linkages among groups conducting research, government bodies responsible for water policy, and groups representing the agricultural community help to lend credibility to new technologies. Government involvement and support later is commonly required to make widespread implementation a reality through regulatory or legislative changes, education, and, frequently, financial incentives.

Politics and monitoring are also involved in the struggle of resource managers to invoke policy and legislative changes that are more conducive to the ecosystem, or watershed, approach. Because the concept of resource management at the ecosystem level is a relatively new one, managers in many parts of the world are at present engaged in the process of lobbying for policy changes to back up their relatively new way of thinking. Much resource management legislation is reactive, having been drafted to address some historical crisis or conflict over resource exploitation. Consequently, many laws support narrowly focused management of specific resources (trees, deer, fish, etc.) on a piecemeal basis—most often to facilitate maximum growth and therefore maximum harvest.[5]

However, in North America, well-educated stakeholders and residents support a more holistic approach to resource management, and this growing body of support represents political pressure toward legislative reform. One illustration of this political dynamic concerns the permitting of surface water extractors in southern Ontario. Backed by the Ontario Water Resources Act, the Ontario Ministry of Environment regulates water takings by requiring users to obtain a permit for extractions in excess of 50,000 l/day (~12,000 gal/day). Traditionally, Permits to Take Water were granted to proponents so long as extractions did not exceed 10% of the available flow at the site during the base flow period. Stream flow data, however, from drought years (especially in 1998 and 1999) bolstered the already growing concerns of stakeholders who pointed out that the Permit to Take Water Program failed to do the following:

- Recognize the impacts of cumulative takings in catchments experiencing high demand (i.e., multiple users in each subcatchment vying for available base flow for irrigation, municipal, industrial, and other uses)
- Consider the watershed health linkages between water extractions and such things as fisheries, aquatic habitat, recreational, and aesthetic uses

In many areas, pressure from stakeholder groups and Conservation Authorities (lead watershed agencies in Ontario) is leading to a more holistic and subwatershed-based review of Permits to Take Water and has led to the implementation of compliance-monitoring programs.

One obvious and mutually beneficial political linkage among regional politicians, lead watershed groups, stakeholders, and monitoring programs relates to the fact that many politicians use environmental improvements or protection as a fundamental election platform. Most local governments also will (on occasion at least) attempt to reassure their constituents that they are doing everything possible to protect the environment while also promoting community growth and prosperity. For this reason, governments, individual elected officials, and future candidates can all benefit from being seen as supportive of monitoring activities that track watershed health and measure the effectiveness of programs intended to protect it.

We made the claim in Chapter 1 that watershed management plans and the programs contained in them should be established though consensus. The process of consensus building—whether it relates to determining what parameters will form the basis of surveillance and performance-monitoring programs, establishing rehabilitation targets, or other aspects of watershed management—is a highly political process. While engaged in consensus-building negotiations about all aspects of watershed management, members and partners of the lead watershed agency need to know that they have political support from the groups that they represent. Stakeholder groups intent on resolving fisheries issues need to know that they are backed by the wishes of local angling groups, commercial fishermen, outfitters, and so on. Agricultural representatives who wish to ensure that their voices are heard during discussions of base flow, groundwater allocation, and phosphorus reduction programs need backing from producer groups and from consumers of their products. Urban politicians who wish to ensure that there is ample assimilative capacity for future growth of their communities need to know that they are supported by the development community, chamber of commerce, and local residents.

In Ontario, Canada, where most of the populated regions of the province are covered by Conservation Authority jurisdictions, political involvement in monitoring and other watershed management programs is legislated according to a governance model laid out in the Conservation Authorities Act. The decision-making bodies overseeing Conservation Authority business are boards of appointees from municipal governments within the watershed jurisdiction. Most of these appointees are elected officials themselves or are residents with a special interest in watershed management who report directly to their respective town councils. In areas where political involvement is not a legislated necessity, lead watershed groups must solicit political involvement. Some strategies for doing so are discussed in Section 2.2.

## 2.2 Creating political linkages

In Section 2.1, we discussed the need for political support and involvement within the watershed management cycle and gave examples of some common linkages among monitoring, political involvement, and other aspects of watershed management. In this section, we will describe some practical techniques for gaining political support and involvement in watershed health monitoring activities.

In Case Study 1, we relate our experiences in implementing the Closed-Loop Watershed Health Monitoring Model in the Laurel Creek Watershed surrounding Waterloo, Ontario. Because the lead agency responsible for coordinating monitoring activities in that case is the City of Waterloo, strong political involvement ultimately facilitated the recognition of monitoring in the city's operating budget. In addition to offering examples of how to build community education, awareness, and involvement into monitoring programs, the Laurel Creek case study also demonstrates some of the key tenets relating to building and maintaining political linkages between monitoring and the rest of the watershed management cycle. These are summarized below:

- Involve and foster the interest of key decision makers and stakeholders as a means of bolstering political support. Evidence that this philosophy was adhered to in the Laurel Creek Watershed Monitoring Program includes invitation and subsequent representation of all key stakeholders in the monitoring program's steering committee, including senior City of Waterloo staff and members of the city council.

- Earn the support of ambitious, influential, and credible members of the community. In the Laurel Creek example, the steering committee sought representation of key city councillors who would be capable of conveying the importance of the program and its results to the City (a key funding partner). Key traits of participants included competence, credibility, and ambition. The monitoring group sought out politicians for the steering group whom it believed would take ownership of the issues and elevate them such that they could ultimately be used as a platform for re-election or to reach higher positions of political office—decidedly, a mutually beneficial relationship!
- Convey the personal side of monitoring. Beyond inclusion of parameters in the monitoring program that actually measure the human aspects of watershed health (things like population, jobs, and economic indicators), it is also important to directly involve decision makers, stakeholders, and the broader community in the monitoring program wherever possible (see Chapter 4). A significant amount of assessment data generated by the Laurel Creek program was collected by volunteers, as is very common in watershed monitoring projects in the United States. In addition to being very cost effective, collecting stream invertebrates, fish, and water samples instilled a sense of ownership in the data for volunteers, who showed great commitment and gained personal satisfaction from the effort. City of Waterloo council members and staff also accompanied monitoring personnel in the field, providing an intimate setting for the exchange of ideas between watershed decision makers and stakeholders.
- Foster timely and solution-oriented reporting. Monitoring is all about generating information as a basis for decision making. Indeed, it drives the management process. Again, concentrating on the Laurel Creek example, adequate reporting and continual efforts to boost public awareness and education and provide opportunities for participation were viewed as being critical to watershed monitoring efforts. Deliverables of the program included straightforward watershed health "report cards" (analogous to those evaluating a student's progress) that were circulated to the council and all stakeholder groups. Program participants also made frequent appearances before the council to answer questions relating to the health of Laurel Creek and successes of rehabilitation programs. Annual report cards were important awareness and educational tools, but practitioners also saw them as opportunities for adaptation in the management cycle. They planned seminars and workshops that coincided with issuance of the report card to help bridge the gap between measuring the status and developing solutions to problems.

- Show the value of the program in terms that political decision makers can relate to, and concentrate on the positive. In the Laurel Creek example, reporting was seen as a strategic exercise; rather than simply reporting on environmental indicators or population growth, monitoring reports were used to maintain and build on program support to enable long-term stability of the program. Reports concentrated on progress. Rather than reporting that phosphorus levels in Laurel Creek were higher than targeted levels, reports showed a slow but steady decrease in percentage exceedance of the target and described what this meant in terms of growth potential (increased assimilative capacity) and recreational opportunities for the community.
- Ensure that programs are cost effective. In reality, monitoring is often perceived as being of secondary importance to other management activities that are directly responsible for progress toward goals, for example, flood warning, rehabilitation, and growth. Although dispelling this perception is one of the tasks of the education and reporting components of the monitoring program, it is also prudent to ensure that the costs of actual data collection and analysis are reasonable within the budgetary context of overall watershed management.

## 2.3 *Avoiding pitfalls*

Understanding political nuances is never easy. There are countless factors that affect the way that people view issues and interact with one another. Personal biases, preconceived notions, personality conflicts, ignorance, and many other factors can hinder our ability to forge the political linkages that are so important to the monitoring program. Whereas the Laurel Creek examples in Section 2.2 assisted us in outlining several general strategies for bolstering political support, this section will describe some specific hurdles that practitioners often face and will suggest ways to overcome these barriers.

Some of the most troublesome situations that practitioners will have to deal with in the process of generating political support for monitoring activities include the following:

- Succession of political figures (elections, reappointments, etc.)
- Reporting of disappointing results
- Building consensus
- Poor fit between short-term political time frames and the long-term process of performance evaluation

By definition, the succession of leaders is inevitable in the political forum. Political succession may be seen in a positive light when strictly personal differences between stakeholders cause stalemates in the consensus-building process. Equally often, though, significant setbacks can occur when knowledgeable and committed supporters are relieved from their posts. For this reason, partnership- and support-building exercises must always consider *succession management.* Because most monitoring programs are directed by some form of steering committee (often a multi-stakeholder group that reports to the lead watershed agency), one simple technique that we have seen used to assist in maintaining consistency is to specify automatic invitations for past chairs or presidents to remain within the general membership of the steering committee for some time after they have been succeeded in office. We have discussed the idea of involving all stakeholders on steering committees, and although it is important that committee size does not become unwieldy, the effects of succession can be minimized if stakeholders can send more than one representative. Here, there is another obvious tie-in between reporting and educational or awareness functions of the monitoring program: the better educated the members of the community and stakeholder groups that the steering committee represents, the easier the transition when new representatives come forth.

As illustrated in Figure 1.2 and discussed in detail in Chapter 3, surveillance and performance evaluation are both critical components of the watershed management cycle. But the backlash that can result when practitioners are forced to report disappointing results can be debilitating. In the case of surveillance data, disappointing results may include such things as declining water quality trends over time or space, exceedances of pollutant criteria in surface or groundwaters, and many others. Although this type of information can certainly cause some unrest in the political arena, it often can exert a positive influence on overall watershed management by boosting support for rehabilitation efforts. The main strategy here is to make sure that reports are solution oriented, in other words, are tied into other programs within the management framework that can alleviate the problem. This is one situation in which the reactive tendencies of the political process can be positive because monitoring data may act as a powerful catalyst for new programs. Performance evaluation data (i.e., data reporting on the successes of programs themselves) that are disappointing can cause an equivalent amount of political upheaval, and because the upheaval is brought about by results of the actions of specific people and programs rather than merely by the status of resources, it can be more difficult to mold political reaction into positive action. From the point of view of those involved in the monitoring program, a few strategies can be very helpful. Before delivering reports to decision makers, it is always prudent to discuss monitoring results with those involved in the evaluated program to determine whether any extenuating circumstances contributed to shortfalls in delivery and to discuss how the program can be improved. When subsequently reporting the data to a broader audience, stress that the data will be used to improve programs and that, by its very nature, management is an adaptive and iterative process. Maintaining this positive emphasis is not easy to do, but politicians need to understand that because of the complexities associated with watershed management, activities do not always work as well or as quickly as planned. Focus on progress. Although we do not advocate wanton "spin doctoring," it is important in the political arena to focus on the positive. Even if the best that can be made of an unsuccessful initiative is to learn from one's mistakes, this should still be seen as progress.

Consensus building can be a difficult and time-consuming political process. The effort often includes members of monitoring teams (who can report on watershed health data and describe which management efforts have worked and which have not) as well as participants in other programs within the management framework. The dichotomy here is that although representation by all stakeholders with an interest in a particular watershed issue is desirable for the most informed decision making, the more opposing views are represented, the more difficult it is to achieve consensus. One significant hurdle to overcome in the consensus-building process relates to the typically antagonistic political process to which practitioners are normally exposed. The political norm is to solicit overwhelming support for one ideal so as to exclude opposing interests. One of the most difficult tasks associated with initiation of the consensus-building process is to shift the mind-set of participants in favor of a more cooperative, teamwork approach. The fundamental purpose of consensus building is to permit information exchange and open negotiation between various stakeholders and to prevent alienation of any one stakeholder with markedly opposing views from the rest of the group. The first step is to ensure that everyone involved is cognizant of the overall goal of watershed management and understands that maintaining watershed health means maintaining balance between human and ecological needs. It is helpful to view contrasting ideas as part of a continuum rather than as mutually exclusive. Recognizing interdependencies between different camps also assists in the development of a team atmosphere.[14,16] When working toward consensus, it is the role of the lead watershed agency to encourage active listening, clear expression of ideas, constructive and creative critique of options, visualization, and reforming of situations.[17]

Another common pitfall relates to the fact the long time frames associated with surveillance and (particularly) with performance evaluation rarely mesh with the short-term agendas in the political arena. One of the best ways to address this issue in relation to performance evaluation is to build a system of milestones into the planning process. One hypothetical example of a specific, measurable, attainable, relevant, and time-bound (SMART) target may be "to provide critical forest habitat for upland species, promote groundwater recharge, and supply lumber for fuel and lumber by establishing 30% watershed forest cover by 2050." As one might suspect, generating the support for a long-term reforestation program that would be needed to achieve the above target may be difficult to do among political leaders. However, support may be gained if the overall target can be broken down into shorter-term milestones, which may include such things as employment and economic spin-offs associated with planting 1,000,000 trees per year, operating infrastructure like seedling nurseries and tree distribution networks, creating a landowner education and extension service to generate planting sites, and so on.

# Scientific assessments

*Nothing in nature is isolated. Nothing is without reference to something else. Nothing achieves meaning apart from that which neighbours it.*

**—Johann Wolfgang von Goethe**[18]

Scientific assessment is the second component of our Closed-Loop Watershed Health Monitoring Model. In this chapter, we describe why it is an important component of any successful monitoring program. We describe some techniques for performance evaluation and propose a framework for surveillance monitoring (Section 3.3) that enables practitioners to stratify sampling effort according to the types and severity of impairments likely to be found in watersheds based on their size and land use patterns. We also deal with data quality assurance (Section 3.4) as well as with geographic information systems (GIS) and modeling tools (Section 3.5).

Watershed health monitoring should evaluate the status of cultural and socioeconomic attributes of a watershed, including recreational opportunities, assimilative capacity, safety and availability of drinking water, aesthetics of the natural environment, population growth, tourism, public health, and employment. Because the focus of this book is on aquatic aspects of watershed health monitoring, examples in this chapter will concentrate on techniques for assessing *aquatic resources.* Our intent, however, is to provide a set of techniques that are transferable (i.e., can be adapted for use in monitoring other aspects of watershed health).

## 3.1 Rationale

Scientific assessment is what immediately comes to mind when most people think about monitoring of any kind, and there are a number of reasons that it appears in our Closed-Loop Model (Figure 1.3). Obviously, scientific assessment provides us with information on the status of watershed health and how it is changing over time. Surveillance data (as discussed in the next section) also provide a basis for setting targets during the watershed-planning process. Later in the management cycle, when programs are implemented to protect and restore ecological function or to develop and promote human uses of watershed resources, performance evaluation becomes an important component of scientific assessment because it measures progress toward watershed management targets.

The importance of scientific assessment is highlighted by work such as the U.S. Environmental Protection Agency's (EPA's) *A Watershed Approach Framework,*[12] which discussed the need for "strong science and data" to support an iterative watershed management system, with data collection required for the following:

- "Assessment and characterization of ... natural resources and the communities that depend on them"
- Target setting and development of action plans based on the health of the aquatic system and needs of the community (to support watershed studies and the watershed-planning process)
- Evaluation of the effectiveness of management decisions and revision of targets and management plans (performance evaluation)

Paralleling discussions in other chapters, the importance of scientific assessment is further reinforced because it is inseparably linked with the other components of the Closed-Loop Model. It is necessary for the development of educational tools and forms the basis of a watershed health message that ultimately wins over community and political support, thereby justifying commitment to long-term funding.

## 3.2 Surveillance and performance evaluation

Surveillance and performance evaluation are the two components of scientific assessment. Figure 3.1 contrasts these two genres of monitoring based on their relationships to watershed management.

| | Performance evaluation | Surveillance |
|---|---|---|
| **Purpose:** | Determine how effective implemented management programs are | Evaluate environmental quality and change over time and space |
| **Measures:** | Administrative actions (e.g., field projects) | Indicators of environmental quality |
| **Role within adaptive management:** | Improve programs, modify targets, etc. | Prioritize administrative actions, identify watershed health issues, support development of goals and targets. |

*Figure 3.1* Contrasts between performance evaluation and surveillance monitoring.

As illustrated in Figure 3.1, surveillance data allow us to answer questions such as the following:[19–21]

- What is the condition of resources within the watershed (both aquatic and other)?
- How and why are stream health attributes changing over time and space?

In doing so, surveillance provides a basis for program development and prioritization (e.g., development of water quality criteria and rehabilitation programs). Performance evaluation, on the other hand, provides the management cycle (Figure 1.2) with its capacity for adaptation. It empowers us to learn from our successes and mistakes and to adjust our strategies accordingly (see Figure 3.2).

## 3.3 The three-tiered surface water assessment model

In this section, we propose a model (Figure 3.3) that provides direction on how to answer the subset of surveillance questions related to the assessment of surface water. The purpose of the model is to permit sampling effort (and

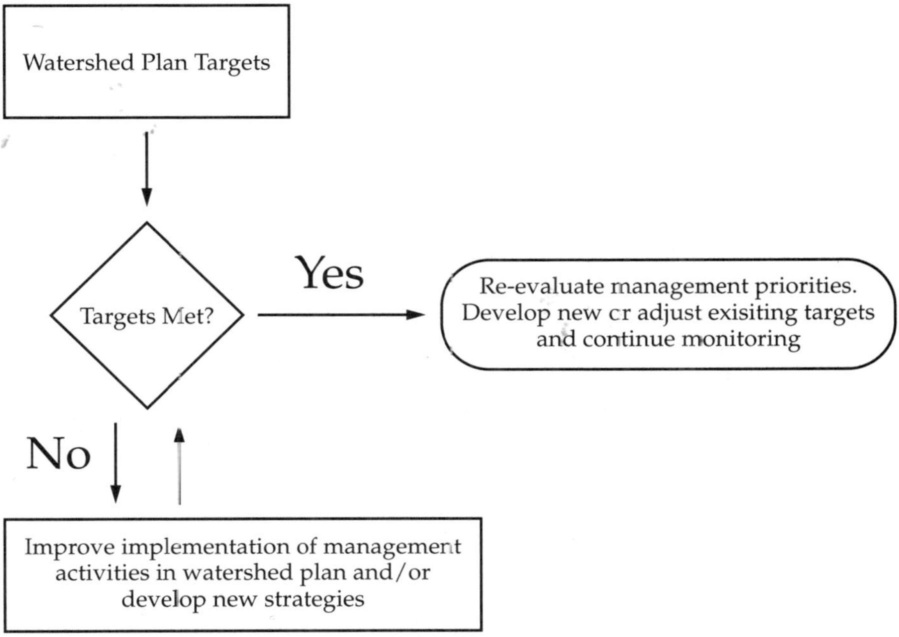


***Figure 3.2*** The relationship between performance evaluation and surveillance in adaptive watershed management.

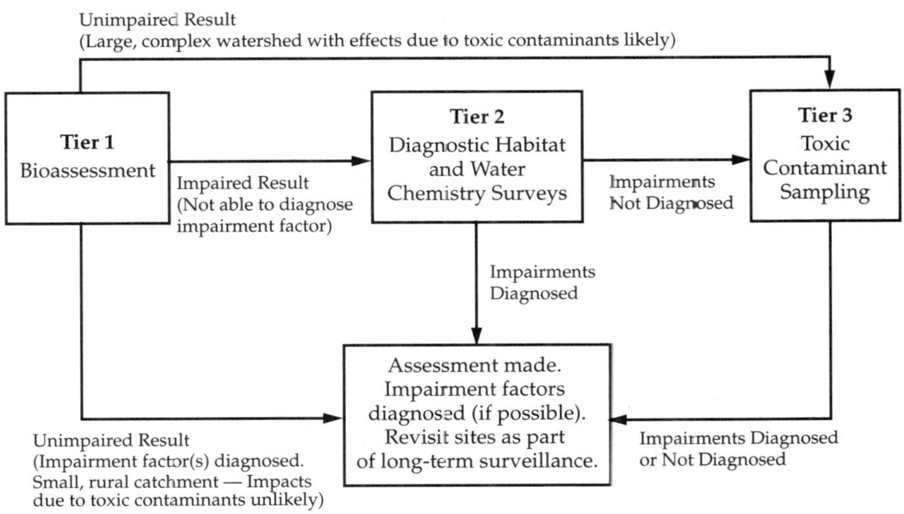


***Figure 3.3*** The three-tiered surface water assessment model.

the allocation of associated personnel and financial resources) to be stratified according to land use in the catchment (which we assume here to be a good predictor of the probability and severity of various water chemistry, habitat, and toxic impairments).

Biological assessment (see Section 3.3.2) is specified as the first step in our model, reflecting our biological definition of aquatic health. Recognizing that equally impacted sites from similar locations can have markedly different biological communities, we define the status of surface water health as a measure of the similarity between the aquatic community found at a test area and that found at unimpaired and desirable reference sites with similar physiographic potential. Test sites with communities that match well with reference communities are considered unimpaired (natural processes are the main factor shaping the living aquatic community), whereas test sites that show marked deviation from the expected condition are considered impaired (human influences are primarily responsible for community composition). In many areas—for example, unimpaired sites in rural watersheds or impaired sites for which the chosen biological assessment tool can adequately diagnose the impairment factors—biological approaches may be the only assessment tools required.

Tier 2 (water chemistry and habitat surveys) and Tier 3 (toxic contaminant sampling) approaches augment the diagnostic abilities of bioassessments. These approaches are most important in watersheds where land use suggests that test areas are (or are likely to be) impaired by a suite of impairment factors. Yoder[21] proposed a similar framework and stressed that biological (i.e., Tier 1) approaches should base assessment on data from at least two aquatic communities, such as fish, macroinvertebrates, or periphyton. He further asserted that practitioners should consider stream morphology, discharge, substrate, riparian habitat, and various water quality indicators to assist in diagnosis of biological impairments. Yoder[21] also stressed the need for other management strategy–specific monitoring activities to round out a watershed health monitoring strategy, specifying surrogates for recreational quality (e.g., fecal bacteria, other pathogens, and organics), safety of the consumption of fish, and others. Similarly, Barbour et al.[22] stressed the importance of combining biological assessments with water chemistry and habitat surveys to facilitate the identification of specific stresses in impaired streams; confirm the locations of point source impacts; develop appropriate discharge limits and treatment facilities to meet these limits; and monitor compliance.

### 3.3.1 *Model assumptions*

A successful assessment framework must recognize the applications and limitations of available assessment tools and arrange them in a hierarchy that reflects their complementary nature. At the same time, the framework must recognize that an integrated approach to monitoring aquatic resources is important if surveillance is both to distinguish impaired from unimpaired sites and to establish causation. The hierarchical structure of our model recognizes the contrasting strengths of biological assessment tools and chemical and habitat surveys,[21] as explained below.

Water chemistry–based approaches seek to measure individual parameters (usually pollutants) and are usually related to specific water quality criteria or pollution abatement policies. They take a surrogate approach, whereby measured parameters are assumed to reflect some ecological response; for example, suspended solids may be measured as a surrogate for enrichment or as a measure of soil erosion within a catchment. So relying on water chemistry data alone is a problem for resource managers who may not be able to measure all of the parameters that account for the differences between "healthy" and "unhealthy" sites. The approach is particularly problematic when synergistic or antagonistic mechanisms are at work. According to Norris,[23]

> Physical and chemical variables are mostly compared with guidelines that are set either as "best guesses" by expert opinion or based on the results of toxicity tests. Guidelines are commonly applied in a blanket fashion within political regions. They usually do not account for possibly varying effects in different environments.

Biological approaches are complementary because they are *direct* assessment approaches. Rather than measuring some surrogate for ecological function, the ecological response is directly measured. Biological approaches, especially those using benthic invertebrates, fish, and periphyton as indicators, have been used for years in studies seeking to assess the health of all types of aquatic habitats[22,24] and are widely considered to be good indicators of aquatic ecosystem health.[22,25–36] Use of the organisms inhabiting water bodies as indicators is often thought of as a holistic means of assessing ecological impact because the pattern of biotic distribution is a manifestation of all the human and natural conditions to which biota have been exposed over their life cycles.

Water chemistry approaches tend to give partial coverage in time and space because the data reflect instantaneous conditions at the collection site.[21] They are reflective of a bottom-up approach, whereby managers seek to synthesize, predict, or evaluate ecosystem-level effects based on individual components. Chemical approaches, therefore, are most useful in specifying design criteria for projects but often leave managers wondering whether results are biologically meaningful. Bioassessment approaches—which, in isolation, often leave managers with many reductive questions about exactly what land–use interactions are causing biological differences between sites—blend with water chemistry monitoring programs because they measure cumulative effects. Rather than measuring surrogates, they evaluate biological responses to the full rigor of the aquatic environment and, as such, are distinctly a top-down approach.[21,22,37] Biological approaches provide complete coverage[21] because unlike snapshot-like water-quality assessments, bioassessments reflect conditions over time (i.e., over the life cycles of the organisms studied) and space (because communities studied at one site have been subjected to water quality and habitat conditions that are impacted by upstream areas of the catchment). Although water quality–based approaches are well suited as design criteria (e.g., for specifying maximum pollutant levels in effluent), natural variability among equally healthy sites generally precludes the use of biological approaches in this manner. Bioassessment approaches tend to be resource focused rather than pollutant focused; hence, they are better suited as impact assessment tools. Biological assessments (particularly since *rapid bioassessment protocols* have been available) can be more cost effective than chemistry or toxic contaminant sampling procedures.[20,22,38] Where specific health standards do not exist, for example, in the case of rural areas, with diffuse water chemistry and habitat degradations, biological assessments may be the "only practical means of evaluation."[22]

Beyond the strengths of biological approaches as assessment tools, further justification for their prominence in our model owes to the fact that most environmental legislation, for example, the Ontario Water Resources Act[39] and the U.S. Clean Water Act,[40] stresses the protection of human uses of biota (and occasionally the biota themselves). Biological assessment data support many restorative programs and regulatory frameworks, such as the Total Maximum Daily Load (TMDL) Process in the United States, and can aid in developing standards for various pollutants according to biotic responses.[22] Government agencies (particularly in North America) have endorsed biological assessment techniques by funding their use in large-scale programs.[34]

Although providing general direction on implementing an assessment hierarchy but not specifying indicators or sampling methodologies, our assessment model recognizes that considerable plasticity is required to enable monitoring activities to respond to a range of watershed issues; allow various interpretations of watershed health; and recognize the limitations associated with funding and expertise.

Another consideration for our assessment model was that it must remain valid regardless of the scale implicit in assessment questions. Trends in watershed health (which can be expressed according to Figure 1.1 as shift toward or away from balance) are expressed as changes in surrogate values over time and space. It is critical to understand that our ability to manage watershed health varies in both time and space, as does our approach for assessing health. At smaller scales, for example, a small subcatchment, there may be fewer conflicting human needs and uses of the watershed, as in a small rural and agricultural catchment with several farms. Moving to a larger catchment scale, we may have other industries, rural and urban landowners, foreign ownership, and so on, which introduce complexity on both sides of the watershed health equation. Larger spatial scales, therefore, may require modifications to sampling strategies (perhaps a shift toward greater use of remotely sensed data, a more complex rationalization scheme for locating assessment sites, and a schedule for rotation to allow for complete coverage despite limited funding), but the fundamental tiered approach to assessment is unaffected.

### 3.3.2 *Tier 1: Biological assessment approaches*

Tier 1 bioassessments use patterns of biotic distribution and abundance to assess environmental change across time or space. Remember that the link to Tier 2 monitoring is required to diagnose impairments (establish causative linkages) by correlating biological with physicochemical changes.[41]

Biological assessments can take many forms. The focus may be on individual organisms, species assemblages, or entire communities.[42] The most common bioassessment indicators are benthic macroinvertebrates, fish, and periphyton. As each of the above groups is made up of a multitude of species, wide-ranging habitat and water quality requirements offer a spectrum of possible responses to human perturbations and provide a basis for diagnosing impairments,[22,27,34,41] as discussed below. In the case of the benthos and fish in particular, tolerances of many taxa are well known. For this reason, degraded conditions can often be detected by an experienced biologist with only a cursory examination of the community present at a site.[22]

In relation to benthic invertebrates, Rosenberg and Resh[34] indicate that they "act as continuous monitors of the water they inhabit, enabling long-term analysis of both regular and intermittent discharges, variable concentrations of pollutants, single or multiple pollutants, and even synergistic or antagonistic effects." Biota are abundant in all types of aquatic systems, living on or in the substrate or in the water column, and therefore can be used to evaluate environmental health in all types of aquatic habitats. There are many accepted methods for collection and analysis, and in many cases (particularly for benthos and periphyton), only relatively inexpensive gear is required. Adequate dichotomous keys permit identification of most of the biota commonly used as indicators; however, some groups can be problematic, such as midges (Chironomidae) and aquatic worms (Oligochaeta) within the benthos. With respect to periphyton, several nontaxonomic approaches that characterize algal assemblages based on functional and structural characters (i.e., using biomass or chlorophyll measurements) are available.[22] In contrast to most fishes, benthic invertebrates and periphyton are relatively sedentary and remain in localized areas throughout their life cycles (at least the aquatic stages). The life cycles of periphyton are quite short, making them ideal for studies investigating intermittent impacts and their duration.[22] Fish and benthic macroinvertebrates, with life spans that range from weeks to years, are good integrators of medium- to long-term environmental impacts. Because of the restricted mobility and habitat preferences associated with the benthos, they are "subjected to the full rigor of the local environment"[27] and are often considered preferable indicators for studies looking at spatial and temporal disturbance patterns at the site scale.[34] Fish, with relatively long life cycles and expanded home ranges, may be better indicators of wider ranging habitat and water quality conditions.[22]

Although the periphyton, benthos, and fish groups have differing strengths as indicators, periphyton and benthic invertebrates provide additional advantages because they can be used as early warning systems for managing fisheries (they may respond more quickly to perturbation, and their collection does not directly impact the resource). They can also be used successfully in smaller systems that generally support diverse communities of periphyton and benthos but only a limited fish fauna.[22] Fish, by comparison, are distinctly the group of choice for characterizing fisheries resources directly, and many policy designations and regulations are based on their community composition (i.e., "cold water" and "warm water" designations that are often closely meshed with regulatory requirements and protection). Each of the groups also represents an important source of biodiversity, including threatened species. Fish, for example, "account for nearly half of the endangered vertebrate species … in the United States."[22]

To this point, we have focused on the strengths of bioassessments and the various groups of indicators used. To maintain a balanced treatment, we next discuss some of the challenges associated with bioassessments. Further justification for Yoder's[21] exhortation regarding the need to use more than one type of biological indicator is in the fact that no single category of indicator will necessarily exhibit responses to all types of perturbation. Benthic invertebrate communities, for example, may show no response to low levels of herbicide pollution that are nonetheless high enough to disrupt downstream plant communities.[34,43] Because of their trophic status as primary producers, periphyton are most directly affected by physical and chemical factors. Invertebrates and fish, with increased mobility, longer life cycles, and an often greater range of foraging options, theoretically are less directly affected.[22] Other problems relate to the fact that "healthy is variable" or that "no single definition of a healthy biological community is appropriate."[44] Biota exhibit "spotty" or "contagious" distributions,[34] often necessitating replication to produce (and driving up costs associated with) precise density estimates used in assessment methodologies.[34,44] We should note though, that subsampling techniques and rapid bioassessment approaches[22] can relieve this problem to some degree.

Other difficulties relate to the fact that many factors other than water quality—such as seasonal patterns (especially among the insects), substrates, current velocity, and temperature—can affect the distribution and abundance of organisms just as human perturbation can. For this reason, careful design of studies and knowledge of the ecology of the indicator biota must be brought to bear in bioassessment programs.[34] We advise a cautious approach to practitioners planning to use a metric or index-based approach to data analysis (as discussed in Section 3.3.2.1), because the large number of biotic and diversity indices available in the literature (particularly relating to benthic invertebrate assessments) is at least partly due to limitations in diagnostic ability and other difficulties.[34,41]

Adequate designs of bioassessment studies—which take into account the strengths and weaknesses of each indicator group—will allow practitioners to assess the effectiveness of management activities (evaluate performance); develop predictive models (often useful for generating hypothesis-of-effect pathways as part of the process of developing or refining management strategies); and diagnose impairment factors. To do so, designs must facilitate comparisons by doing the following:

- Specifying collection of relevant data[44]
- Eliminating sources of error (i.e., confounding effects) that reduce our ability to distinguish real differences between sites or times from natural variability[41]
- Applying suitable analytical procedures[41]

As there are many discussions in the literature regarding sampling methodologies for bioassessments,[22,26,27,35–37,44,45] our discussion in the following sections will focus on analytical techniques—an area in which substantial advances have been made in recent times but significant discord still exists within the scientific community.

#### *3.3.2.1 Univariate approaches*

Analytical procedures for biological assessments fall into two main categories, univariate and multivariate approaches. Univariate approaches are characterized by the use of indices that express ecosystem health status as a single value.[41] Examples of common univariate indicators include simple abundance, richness, or biomass measures. Diversity and biotic indices that have been used extensively in North America, Europe, and other parts of the world are also part of this group. Related to benthic invertebrate approaches, Resh and McElravy[44] reported that of a collection of 90 published lake and river bioassessment papers, 40% used univariate analytical procedures. According to Resh and McElravy[44] and Norris and Georges,[41] the most common diversity measures used in lake studies included the Shannon[46] and percent similarity[47] indices. Stream studies favored the Simpson Diversity Index. The following is the calculation equation for the Shannon index:[48]

$$H = -\sum_{i=1}^{S} p_i \ln(p_i) \tag{3.1}$$

where $S$ is the number of taxa and $p_i$ is the proportion of the total abundance made up of the $i$th taxon. Shannon's index is based on information theory[46] and assumes that greatest health is manifested as an even distribution of taxa.[41,49] Simpson's index (Equation 3.2) requires counts to be made of total individuals per sample and counts for each taxon in the sample, thus giving a combination of abundance and richness information about a site. It is as follows:

$$D = \frac{1}{\sum_{i=1}^{S} p_i^2} \tag{3.2}$$

where $S$ is the number of taxa present and $p_i$ is the proportion of the total abundance made up of the $i$th taxon. Simpson observed that healthy communities tend to have high richness and that abundance is split relatively evenly among the taxonomic groups present. Interpretation of his index is based on the assumption that the probability of any two randomly selected organisms belonging to a single taxonomic group would be lower for a healthy (i.e., more diverse) community than for a less healthy (less diverse) community.

Biotic indices are another class of univariate indices. They characterize aquatic health based on empirically derived biotic responses to various disturbance regimes and have also been used extensively, particularly in riverine assessments. In most cases, biotic indices were developed to assess community health in relation to a single disturbance regime, such as organic enrichment, acidification, metal contamination, or sediment loading.[42] One example of a biotic index commonly used in Ontario is the Griffiths[27,37] Biological Monitoring and Assessment Protocol (BioMAP) WQI biotic index (see Equation 3.3). The BioMAP index is uncharacteristic because rather than rating taxa responses to a specific disturbance regime, Griffiths claims that the BioMAP index measures "upstream shifts" in the community resulting from many common water chemistry and habitat stressors. The calculation formula is as follows:[27,37]

$$\mathrm{WQI} = \frac{\sum_{i=1}^{n} (e^{sv_i} \times \ln(x_i + 1))}{\sum_{i=1}^{n} \ln(x_i + 1)} \tag{3.3}$$

where $sv_i$ is the sensitivity value of the $i$th taxon, $x_i$ is the sample abundance of the $i$th taxon, and $n$ is the total number of taxa (richness) found in the sample. Griffiths[27,37] asserts that the index is based on the *river continuum concept*,[50] with *sensitivity values* assigned according to where, in unimpaired stream systems, each taxon reaches its highest density (i.e., in small creeks, larger streams, rivers, large rivers, or lakes). Thresholds for index values allow comparisons to be made between a sampled community and a theoretically expected community based on stream size.[20,27,37] Deviation from this expectation (e.g., river dwellers dominating a creek environment) is considered to reflect impaired conditions.

Multimetric approaches, such as the popular Index of Biotic Integrity methodology,[51–53] which uses fish (or other) community measures to quantify biological integrity, are included in this category because they employ summations of individual metrics to derive a final score. The score is then compared against established thresholds (based on physical stream characteristics and geographic location) to assess health.

Biotic and diversity indices have been used extensively because they do the following:

- Assess health based on the concepts of prevalent ecological theories such as the diversity and stability hypothesis and information theory, the river continuum concept, or competitive interaction[27,37,41,46,49,50,54,55]
- Enable minimally trained practitioners to manipulate complex community data using simple mathematical calculations; assessments can be as simple as comparing index values against established thresholds or standards[41,56,57]
- Enable spatial or temporal comparisons of biological data collected using different sample sizes and collection methods[41]
- Are seen as being less costly than multivariate and reference condition approaches (see Section 3.3.2.2) that typically involve sampling at multiple control (reference) sites and collection of numerous replicate samples[41]

Despite their wide application by practitioners, univariate approaches have distinct limitations, and it pays to be aware of the limitations and assumptions associated with each index. In the case of diversity indices, the ecological theory on which the index is based must account for most of the variability between sites or times if the index is to provide an acceptable basis for assessment. Will the possibility of members of the same taxon exhibiting a clumped distribution (for reasons of microhabitat, breeding, or behavior) limit usefulness of Simpson's index in relation to a specific assessment question, for example?[41] Remember also that ecological theories do not hold up in all circumstances. The applicability of Shannon's index (which measures evenness of taxa distribution), for example, is called into question by contradictory evidence that communities often exhibit log-normal distributions.[41,49,58] Furthermore, it has been shown that Shannon's and other indices may be subject to confounding factors associated with clumping and microhabitat effects, as well as implications associated with vastly different body sizes.[54,41] Hughes[59] cautioned that water quality impacts were just one of a number of factors affecting the value of diversity indices and that study design must take into account influences of sampling methodology (size, time of year, etc.) and level of taxonomic detail. Practitioners must also be mindful of the implications associated with comparing sites sampled using different collection methods because index values are often influenced by sample size and habitat type.

Biotic indices too have their own set of limitations. The BioMAP index discussed above has been the subject of much debate recently in the literature[60,61] because of a lack of agreement regarding the following:

- The defensibility of the process by which sensitivity values were assigned to taxa
- The assumption that BioMAP index values decline in a downstream direction in unimpaired stream systems
- The assumption that animals generally residing in larger streams are less sensitive to human disturbances than animals characteristic of smaller, headwater creeks
- Whether impairment thresholds that decline with stream size make it increasingly difficult to detect impairment as systems increase in size
- The applicability of decision thresholds based on bank full-channel width, particularly in relation to atypical streams, for example, streams originating in low-gradient systems with wide full-bank width and fine sediments

As we alluded to above, the fundamental argument against the use of univariate approaches is that condensing biological community information into a single value necessarily results in the (often subtle) trends in community composition being masked.[41] Furthermore, changes in biotic indices and other derived and composite measures of biological health may be seen as containing little information because they do not always respond in a predictable or linear fashion along disturbance gradients and may not show responses to all of the disturbances that are likely to occur at test sites located in watersheds with complex land use matrices.[41]

To summarize our discussion of univariate approaches, regardless of which indices are used, be cautious! Careful consideration is warranted to ensure that collected data will meet the assumptions of the chosen indices and that the chosen indices will effectively assess impacts associated with disturbances that are likely in a given aquatic system according to adjacent land use. In complex catchments, avoid basing assessment activities on a biotic index—such as Hilsenhoff's index[62] — that is calibrated against a single type of disturbance. Take caution when applying an index outside of its area of development. Many indices were developed for a specific area and impact and may only be locally applicable.[41] Consider univariate indices as prescreening tools. They are usually seen as one part of an expert system that requires considerable biological expertise and integrates data from other sources to build a prognosis about health of the system.[41] Undertake exploratory work to establish that the chosen index has biological meaning and reflects biological responses to disturbances relevant in a given watershed (see Section 3.4).

#### *3.3.2.2 Multivariate statistics and the reference condition approach*

The reference condition approach overcomes many of the shortfalls of univariate approaches by accounting for the natural variability among equally healthy sites and eliminating the need to express biological condition as a single value. Indeed, a number of authors believe it to be the most scientifically sound means of assessment available.[23,41,63–65] The assumption here is that—in the absence of human disturbance—the communities at test sites will be similar to those at reference sites with similar habitat conditions.[23] In simple terms, the sampling and analytical procedures involved in applying the reference condition approach involve the following:[23,41,65–69]

- Site selection and sampling to characterize the reference condition
- Classifying reference sites into subsets with relatively homogeneous biological communities
- Matching taxa assemblages with the physicochemical characteristics within biotic groupings
- Selecting the most appropriate groupings of reference sites to generate expectations of community structure that should occur at test sites in the absence of anthropogenic stresses
- Testing for significance of discrepancies between test site community structure and the reference condition to detect impairment or trends over time and space

Central to the steps above are a group of statistical approaches in which each taxa present is considered to be a variable and presence or absence or abundance is considered an attribute of a site or time.[70]

The primary limitation of the reference condition technique is the need for relatively sophisticated statistical analyses and a correspondingly high level of practitioner expertise to ensure that the assumptions implicit in statistical tests are upheld.[41,71] In addition, because replication is done at the site level rather than at the within-site level, and test sites cannot be evaluated in isolation, the technique is often seen as more expensive than univariate approaches. Also problematic is the fact that deviations (which are considered reflective of impairment) that form the reference condition are not necessarily biologically meaningful. Choose your reference sites carefully and use good judgment when interpreting statistics.

### 3.3.3 Tier 2: Water chemistry and habitat assessment

Although bioassessment techniques represent the best choice for distinguishing impaired or undesirable from unimpaired or desirable aquatic environments (Figure 3.1), establishing causation frequently necessitates the collection of abiotic habitat and water chemistry information.[20,22,72] This is true for several reasons. First, correlative or experimental approaches used to diagnose impairment types generally involve partitioning the variation between sites according to changes in physicochemical variables associated with human influence (refer to Section 3.5). Second, most of the best management practices available to us were developed through a bottom-up methodology in which one environmental variable is manipulated to generate some biological response; for example, phosphorus may be limited to control growth of nuisance algae in streams. Finally, physicochemical attributes associated with terrestrial (upland, riparian, etc.) and aquatic (lentic and lotic) compartments of the watershed are major determinants of aquatic community potential[22] and are required to support site selection when defining the reference condition for a test site.

The following sections discuss habitat and water chemistry assessments.

#### 3.3.3.1 *Water chemistry approaches*

As we alluded to in the previous section, water chemistry surveys are frequently carried out to do the following:

- Facilitate site selection as part of a reference condition approach
- Establish causation of biological impairments through correlative analyses and experimentation
- Compare ambient water quality conditions within the watershed against established standards for health
- Allow calibration of predictive models used in watershed-planning exercises
- Evaluate performance of reduction programs for specific pollutants, such as phosphorus and sediment

Water chemistry studies investigating causation of biological impairments must seek to overcome autocorrelative linkages between chemical and habitat variables (as discussed in Section 3.3.5) and the often-dramatic fluctuations in chemical concentrations that occur, particularly in surface waters. In compliance-monitoring studies, in which ambient water chemistry attributes are compared against established standards, the problem of variance in means is problematic (and is further justification for use of biological methods as the first line in assessments of aquatic health). Water quality data are frequently reported for sites using box plots that illustrate means and quartiles within the range of observed values. Because of variability and the bottom-up nature of water quality assessments, stream health risk is often difficult to quantify, and the predictive power of any single parameter tends to be weak. For this reason, composite water quality indices have been used as standards in many jurisdictions. In British Columbia, for example, a water quality index was established that gives a summation of water chemistry scores to report on overall water quality. The summation included the number of water chemistry variables covered in provincial standards that do not meet objectives; the frequency with which objectives are not met; and the amount by which objectives are not met.[73] In the United States, a TMDL compliance framework is used. The focus of the program is to provide a surveillance and performance evaluation function by reporting on the "maximum amount of a pollutant that a water body can receive and still meet water quality standards."[74] Standards are set by states, territories, and tribes at levels to protect various uses of aquatic resources, including "drinking water supply, contact recreation (swimming), and aquatic life."[74]

Table 3.1 provides some examples of common water chemistry parameters measured in correlative, experimental, and compliance-monitoring

***Table 3.1*** Selected Water Chemistry Parameters That Are Commonly Investigated as Part of Aquatic Surveillance Monitoring Programs with Generalized Linkages to the Health of Water Bodies

| Water chemistry parameter | Stream health implications |
|---|---|
| Phosphorus (total P, phosphate, etc.) | Generally the limiting nutrient for growth of primary producers in aquatic systems; high levels (*eutrophication*) shift trophic status, increase plant biomass, and reduce water clarity and oxygen availability |
| Chloride | Frequently enters aquatic systems through storm water inputs (road salt) in chlorinated sewage treatment plant effluents and industrial effluents; aquatic organisms exhibit a range of sensitivities to chloride |
| Nitrogen (nitrate, total Kjeldahl N, nitrite, ammonium, etc.) | Nitrogenous products (especially ammonium) can be toxic to a range of aquatic and terrestrial organisms and are a frequent problem in drinking water |
| Suspended solids | High concentrations are frequent impairments to stream habitats; they cause clogging of fish-spawning substrates and dramatic changes to biotic communities |
| Turbidity | A combination of small suspended particles and living organisms in the water column; high turbidity can often indicate eutrophication, erosion, storm water, and other impacts |

studies. We encourage practitioners planning water chemistry studies to research parameters of choice and study design related to Tier 2 water chemistry approaches.[22,74–84]

### 3.3.3.2 *Habitat surveys*

Like chemical data, physical habitat information is often collected as part of correlative or experimental studies aimed at establishing causation of impairments. Alterations to physical habitat can cause a multitude of impairments to biotic communities in both aquatic and terrestrial zones. Many species have specific habitat requirements associated with their various life history stages, so alteration of any one key habitat may lead to extirpation of that species (a concept embraced by Habitat Suitability Indices, or HSI, models that are discussed below). Fisheries management is a good example of an activity that relies on close scrutiny of habitat because it is a primary determinant of productive capacity. Localized habitat alterations in lotic systems can have far-reaching effects on aquatic communities and general health of the aquatic system. Dams, for example, interrupt the natural flow of sediments downstream, interfere with migration of species, alter trophic status both in the impounded area and downstream, and alter temperature and flow regimes.

Methodologies for collecting habitat data can be lumped into several categories:

- Surveys used to measure habitat suitability
- Surveys used to collect physicochemical data as part of a process for establishing causation of impairment
- Compliance monitoring

One common analytical method for habitat assessment involves calculation of HSIs.[85–87] These models are univariate, expert systems in which a summation of scores on individual habitat measures (e.g., food availability, cover, access to spawning habitats, etc.) is done to generate a single value. This single value (HSI score) is commonly used as an estimator of productive capacity of an area in relation to a specific species of wildlife. In the United States, HSI models are available for a multitude of wildlife taxa, including birds, mammals, fish, invertebrates, reptiles, and amphibians.[86] One example of a habitat suitability monitoring tool being developed and applied in Ontario is the Ontario Ministry of Natural Resources (MNR) Stream Assessment Protocol.[85] Like most HSI models, the MNR approach assumes that in aquatic systems habitat conditions are more important determinants of species abundance than are competitive interactions (an assumption that may frequently break down). The MNR protocol is a modular biological and geomorphologic approach that enables predictions to be made about (species-specific) fish productive capacity. Prediction involves comparison of HSI scores against thresholds based on expert and empirical knowledge of how the model correlates with fish production in various reference streams. The data requirements for the model include surrogates for the following:[85]

- Nutrient status (as indicated by benthic macroinvertebrate community)
- Channel structure (amount of fish habitat types and cover)
- Habitat stability
- Sediment transport
- Thermal status
- Unique features (presence of thermal refugia, habitat capable of supporting large adult fish, in-stream cover, etc.)

In general, HSI approaches tend to be more descriptive than predictive[87] and assume the following:

- HSI metrics and decision thresholds are biologically meaningful, in other words, reflect the habitat preferences of the species in question.
- Emphasis (weighting) is placed on metrics that are most significant to the species in question. For example, if the HSI model calculates a straight summation of all metric scores, the assumptions are that all are equally determinant of the ability of the fish species to utilize available habitat and that the parameters that best account for

differences in species abundance and distribution have been incorporated into the model.
- Chemical or biotic interactions are not significantly affecting species abundance or distribution.
- The habitat measures used in the model are measurable and reproducible.

HSI approaches to habitat monitoring are often conducted as a follow-up to biological assessments or correlative analyses that pointed to stream health or productive constraints caused by habitat degradation. HSI models are also commonly used as performance evaluation tools for management programs aimed at protecting or enhancing populations of key species (sport fish, threatened species, etc.) through habitat manipulations (such as creating and improving reproductive habitat, providing migratory routes and linking habitat patches, etc.). Here they are particularly valuable because they can be related directly to habitat milestones and can be reported immediately—even before the biota have responded to changes. They can also be used as predictive tools when designing rehabilitation projects.

In addition to predicting productive capacity or evaluating the merits of restoration work, habitat data are often collected as independent variables when trying to establish causation of biological impairments in water bodies. There are numerous types of habitat information and data collection protocols to satisfy this purpose.[22,85] Unfortunately, the challenges inherent in the selection of habitat measures are equally diverse. One issue is related to problems with defining and measuring (often complex) habitat attributes. For example, monitoring personnel may hypothesize that flow and thermal regimes are important determinants of biological differences between sites. Defining and measuring these attributes, however, may be problematic, and a number of surrogates for each parameter may have to be evaluated. Potential surrogates for flow regime include hydraulic head, time to peak flow after a rain event, bank full width, and so on. Those for thermal regime include annual thermal minima or maxima; water temperature standardized against a certain air temperature (this reading may be taken at any depth in the water column depending on the hypothesis being tested); presence of thermal refugia for cold-water species; rate of evening cooling (overnight thermal recovery); or rate of cumulative warming during sustained heat waves. Practitioners are also commonly faced with the difficult task of selecting habitat measures that are not significantly autocorrelated (see Section 3.3.4.2) and are independent of other determinants of community composition (i.e., water chemistry parameters). An example of distinct autocorrelation between variables is that of current speed and substrate type in streams. Maximum summer temperature and substrate type may intuitively seem independent but, on analysis, may exhibit rather strong autocorrelation in some cases. Development of techniques for measuring habitat attributes in a way that is reproducible is another challenge. What we mean here is that the variance associated with measures of a given habitat surrogate reported by different crews of monitoring personnel at the same site must be minimal if the habitat index is to have significant predictive power.

In-depth discussions of habitat measures are common in the literature,[20,22,76,85,88–111] and practitioners are encouraged to research surrogates for candidate variables for use in predictive models before investing significant resources in data collection.

### 3.3.4 Tier 3: Toxic contaminant sampling

Toxic contaminant sampling is really a subset of water chemistry monitoring and is done for many of the same reasons as for Tier 2 surveys (refer to Sections 3.3.3.1 and 3.3.3.2). There are some justifications, however, for our breaking out this group of protocols into a separate tier within our assessment framework (Figure 3.3). First, these protocols tend to be more expensive in terms of data collection and analysis and require greater (often multidisciplinary) expertise for study design and interpretation. Second, in addition to the normal hierarchical process for diagnosing impairments, we recommend that practitioners conduct toxic contaminant sampling studies in areas that are at risk of toxic effects, even if Tier 1 biological assessments do not indicate impairment.[20] Our reasoning for this is as follows:

- In certain situations (i.e., where biomagnification is occurring), significant biological impairments may exist at trophic levels higher than those sampled in Tier 1 assessments; this is a particular concern in the case of Tier 1 assessments using invertebrate or periphyton indicators.
- Significant human health issues (e.g., edibility of sport fish or contamination of drinking water) may be present but go undetected in Tier 1 assessments.

In following sections, we will discuss some of the toxic contaminant issues in surface waters (Section 3.3.4.1) and give an overview of sampling techniques (Section 3.3.4.2).

#### 3.3.4.1 Common toxic contaminants in surface waters

Although many practitioners are at least partially familiar with common water chemistry parameters such as phosphorus and pH, toxic contaminants tend to be lesser known. For this reason, we will discuss the properties of several broad classes of toxic contaminants in this section.

Toxic contaminants are generally grouped into three categories: organic contaminants, heavy metals, and organometallic compounds. Some of the most common and problematic of the organic contaminants are the organochlorines, a group of synthetic chlorine-containing organic compounds that share the physical properties of poor solubility in water (hydrophobic) but high solubility in fat (lipophilic). They are stable and resistant to biochemical degradation, hence tend to persist for long periods in the environment. Most are volatile enough to allow long-range atmospheric transport. In water, they tend to adsorb to particulates and accumulate in sediments. They commonly bioaccumulate within aquatic biota.[112]

Heavy metals such as mercury and lead are naturally found in water, hence are not considered contaminants unless they occur in concentrations that exceed natural levels. In aquatic environments, metals occur in a variety of chemical forms and biota exhibit a range of sensitivities to each form. In uncontaminated aquatic sediments, metals are usually in mineral complexes or bound to organic matter so are not biologically available. Dissolved metals from riverine or effluent flows generally precipitate in seawater. Bound or precipitated forms are generally nontoxic to organisms; however, transformations such as in methylmetals increase biological availability and toxicity.[112]

A number of attributes associated with selected toxic contaminants commonly found in the Great Lakes and other North American surface waters are discussed in Table 3.2, including the United Nation's "Dirty Dozen": aldrin, chlordane, dichlorodiphenyl-trichloroethane (DDT), dieldrin, dioxins, endrin, furans, heptachlor, hexachlorobenzene, mirex, polychlorinated biphenyls (PCBs), and toxaphene.[113] Jones[20] discussed recent trends in the North American Great Lakes that may help in determining the level of contaminant sampling justified in certain watersheds.

#### 3.3.4.2 *Contaminant sampling issues and techniques*

Contaminant sampling, analysis, and interpretation is complicated by the variety of chemicals in the ecosystem, the complexity of food webs, and differences among water bodies with respect to inputs from the surrounding population. Contaminant levels in fish in the 1970s prompted restrictions on their harvest and consumption. After this, many industries producing or discharging contaminants invested in abatement measures This has led to an overall decrease in contaminant levels (e.g., mercury, DDT, Mirex, and PCBs) in the Great Lakes Basin and other water bodies, especially from identifiable point sources for which progress was made toward reduction or elimination. Unfortunately, ambient contaminant levels in aquatic habitats are not always reflected in biota. This is due to a combination of factors, including changes in ecosystem dynamics due to introductions of exotic species (e.g., zebra mussels) and persistence and biomagnification.

Other complications arise because importation of chemicals into aquatic habitats from nonpoint sources (e.g., atmospheric deposition) can increase ambient concentrations, even once they are banned from point sources within the basin. A number of limnological parameters affect contaminant dynamics, including surface area (of lakes) and catchment areas (of streams), both of which determine exposure to contaminant sources such as atmospheric deposition and outfall of contaminants bound to dust or dissolved in precipitation. These contaminants include organochlorines such as hexachlorobenzene, polychlorinated dibenzodioxins (PCDDs), and dibenzo-furans.[114] These parameters can be used to assess the need for more detailed toxic contaminant sampling strategies.

***Table 3.2*** Human Origins and Uses, Dispersal Pathways, and Toxicities Associated with Selected Contaminants in North American Surface Waters

| Contaminant | Origins and uses | Dispersal pathways | Toxicity |
|---|---|---|---|
| | | **Organochlorines** | |
| Chlordane (a chlorinated hydrocarbon mixture) | An insecticide; all Canadian uses banned in 1995,[112] U.S. Environmental Protection Agency ban in 1988[133] | Volatile—adsorbs to particulates and sediments; persistent and bioaccumulates | Degrades to oxychlordane and heptachlor epoxide (both are more persistent and toxic than most organochlorines); highly toxic to birds[112] |
| Dichlorodiphenyl-trichloroethane (DDT) | Insecticide; Canadian uses banned by 1990,[112] U.S. ban, December 1972,[134] still used in many parts of the world | Volatile and prone to atmospheric transport; its DDE metabolite is commonly bioaccumulated and biomagnifies; DDE and DDD (another metabolite) are more persistent and toxic than DDT[112] | DDE causes eggshell thinning and sterility in fish-eating birds; it is also known to alter hormone and enzyme activity[112] |
| Polychlorodibenzo-dioxins (75 isomers) | Disinfectants and wood preservatives; formed by incineration and manufacture of chlorinated phenols; major sources are pulp mills, waste incinerators, and landfills | Commonly enter waters by atmospheric deposition and tend to concentrate in sediments because of poor water solubility | Range of toxicities; 2,3,7,8-chlorine substitution is highly toxic; implicated in deaths of herring gulls in 1970s; wide-ranging toxic effects in many species[112,135] |
| Dibenzo-furans (PCDFs; 135 isomers) | Most are products of waste incineration[112] | Commonly enter waters by atmospheric deposition and tend to concentrate in sediments because of poor water solubility | Many are endocrine disrupters that affect hormones and reproduction but are less toxic than dioxins; exhibit a range of toxic effects[112,135] |
| Dieldrin | Insecticide; Canadian uses terminated by 1995,[112] banned in United States in 1974[136] | Persistent; bioaccumulates | Toxic to many aquatic organisms; implicated in wildlife mortality (e.g., bald eagles) |

| | | | |
|---|---|---|---|
| Heptachlor | Insecticide; banned in many countries | Highly volatile and is distributed via atmospheric pathways; binds to aquatic sediments and bioaccumulates in fatty tissues[135] | Metabolite (heptachlor epoxide) is persistent and toxic; implicated in impaired reproduction and death in birds and other predators[112] |
| Hexachlorobenzene (HCB) | By-product of industrial chlorination; fungicide; banned in United States and Canada[137] | Long-range atmospheric transport; in aquatic environments, concentrates in sediments; extremely volatile, lipid soluble; bioaccumulates[135] | Enzyme inhibition, immune suppression; can also affect nervous and reproductive systems; carcinogenic[112] |
| Hexachlorocyclohexane (HCH; large group of isomers) | In the United States, lindane (insecticide) production ceased in 1977; production of technical-grade HCH ceased in 1983; lindane is still used in Canada, but other isomers were banned in Canada in the 1970s[138] | Volatile; wide distribution by atmospheric transport; somewhat water soluble | Less toxic than most other organochlorines; reproductive failure and mortality in some birds;[112] can cause "blood disorders, dizziness, headaches, and seizures in people"[138] |
| Mirex | This pesticide was banned in North America in the 1970s; present origins are mainly disposal sites | Persistent; not water soluble or volatile; adsorbs to organics, bioaccumulates; only partly metabolized—slow elimination | Reproductive problems in birds and cancer in lab animals[112,139] |
| Polychlorinated biphenyls (209 individual chemicals) | Many industrial uses because of chemical stability; arise through spills and disposal and incineration | Very stable, persistent; not easily metabolized; bioaccumulates | Carcinogenic; impairs growth, molting, reproduction, and metabolism; in wildlife, causes embryo death, hormonal impairment, and deformities; causes developmental problems in humans |

*(continued)*

***Table 3.2 (continued)*** Human Origins and Uses, Dispersal Pathways, and Toxicities Associated with Selected Contaminants in North American Surface Waters

| Contaminant | Origins and uses | Dispersal pathways | Toxicity |
|---|---|---|---|
| Toxaphene (many polychlorinated camphene isomers) | Broad-spectrum pesticide (insects, fish); banned in North America by the early 1980s[112] | Volatile, long-range atmospheric transport; bioaccumulates[112,135] | Highly toxic to fish; carcinogenic; suspected of causing large bird kills[112,135] |
| | | **Nonchlorinated organics** | |
| Polycyclic aromatic hydrocarbons (PAHs) | Aromatic compounds; main source is incomplete combustion (especially of oils) | Volatility and solubility are variable; concentrated near point sources (urban and industrial); arise as bound particles (e.g., to soot) or via atmospheric deposition; adsorb to particulates, settle out in sediments; persistent; bioaccumulate[112] | Many toxic metabolites, especially in vertebrates; some are carcinogenic, some disrupt epithelia of fish (gills), suppress immune response (clams), and impair reproduction (some marine organisms) |
| | | **Metals and organometals** | |
| Cadmium | Natural sources include eroded soils, volcanoes, and forest fires; industrial sources: mines, smelters, fertilizer plants, thermal electric plants, manufacturing[112] | Metallic form is insoluble; some compounds are soluble; binds to particulates; low pH, hardness, salinity, and particulates enhance mobility; bioaccumulation is variable (modified by pH, temperature, and hardness); does not biomagnify[112] | Causes enemia, retarded gonad development, enlarged joints, scaly skin, liver and kidney damage, reduced growth; toxicity depends on species and age[112] |

| | | | |
|---|---|---|---|
| Lead | Natural sources include weathered rocks, volcanoes, radioactive decay (e.g., uranium); human sources include leaded gasolines (tetraethyl lead), mining and smelting, lead plumbing and solder, sewage sludge, paints, batteries, shot, and sinkers[112] | Dispersal pathways: precipitation, lead dust fallout, storm water, industrial and municipal wastewater; soluble forms are removed by adsorption to particulates and sediments; bioaccumulation varies; organic lead bioaccumulates more and is more toxic than organic forms[112] | Highly toxic (varies according to chemical form, environmental parameters, and biology of species); causes anemia, low growth, liver and kidney damage, and immune and nervous problems[112] |
| Mercury | Sources: chlor-alkali plants (1960s), industrial leachates, smelting, coal burning, release from plants and soil after flooding; especially problematic in large hydroelectric reservoirs[112] | Bacteria convert nontoxic inorganic forms into toxic methylmercury, which bioaccumulates and biomagnifies; rapid accumulation in fish; some animals are able to demethylate and reduce toxicity[112] | Inhibits photosynthesis and growth in phytoplankton; causes nervous system disorders and brain lesions in birds; kidney, liver, and brain accumulation in fish-eating mammals causes neurological and kidney damage, weight loss, and death[112] |
| Organotin | Used in manufacture of polyvinyl chloride; also in catalysts and additives; arises in aquatic systems through production, processing, disposal, and abiotic and biotic chemical pathways[112] | Forms and fate are largely unknown; persistence variable; not volatile from water; bioaccumulation occurs despite biotic and abiotic demethylation[112] | Tributyltin causes shell and reproductive organ abnormalities in shellfish[112] |

Biological resources of a water body can also influence many factors related to concentrations and toxicity of contaminants. Metabolism of DDT, for example, into the more toxic and persistent dichlorodiphenyl ethane (DDE) is a common phenomenon. Similarly, microbes in aquatic sediments often convert inorganic mercury into methylmercury, an appreciably soluble, biologically available organometal that is highly toxic and difficult for most animals to break down. Thus, biotic interactions cannot be ignored in toxic contaminant studies.

Sampling studies related to toxic contaminants usually are undertaken to do one or more of the following:

- Determine how the contaminant is partitioned into biotic and abiotic segments of the environment (Is the contaminant bioaccumulated? Biomagnified? Volatile? Persistent? Water or fat soluble?)
- Document toxicity (through bioassay) and modes of action
- Determine sources of contaminants
- Establish cause–effect linkages

Environment Canada[115] describes recommended methods for "selection of sampling sites within a watershed, and the collection, handling, storage, transportation, and manipulation of samples of whole sediments … for the purposes of physicochemical characterization and/or biological assessment using whole sediments, pore waters or sediment elutriates." The U.S. EPA[116] provides guidelines to regulatory agencies responsible for controlling discharge of toxic contaminants to surface waters. The document may also be useful to practitioners involved in watershed health monitoring activities because it contains information on water quality standards and criteria, effluent characterization, health hazard assessment, and compliance monitoring. The document also contains a case study describing procedures for screening, effluent characterization, exposure assessment, and permit limit derivation. Lee and Jones-Lee[117] suggest sampling protocols for storm water and landfills. Protocols for sediment sampling are given by Mudroch and MacKnight.[118]

#### *3.3.4.3 Multiple regression, a correlative and predictive tool*

Once an appropriate measure of biological health has been selected and used to assess various sites, practitioners are still commonly faced with questions such as the following:

- What environmental variables are contributing the most to differing biological conditions at various sites, or at the same site over time?
- How will biological communities respond to a given rehabilitation technique?

One of the more common means of answering these questions is multiple regression. This statistical technique (see Equation 3.4) extends linear regression to partition variation between sites or times among any number of independent variables, resulting in relatively simple predictive models.[119,120]

$$E(Y|x_1, x_2, \ldots, x_k) = \beta_0 + \beta_1 x_1 + \beta_2 x_2 + \cdots + \beta_k x_k - c \tag{3.4}$$

where $E(Y \mid x_1, x_2, \ldots, x_k)$ represents the expected value of $Y$ given the values of independent variables $x_1, x_2, \ldots, x_k$, $\beta_0$ represents the $x$-intercept coefficient; $\beta_1, \beta_2, \ldots, \beta_k$ represent the regression or slope coefficients associated with $X_1, X_2, \ldots, X_k$ (representing the amount that the dependent variable $Y$ changes per unit change of each respective independent variable, assuming that the value of other variables remains constant); and $c$ is the $y$-intercept coefficient (the expected value when independent variables are 0).

Given the many possible interactions between environmental and socioeconomic variables considered in watershed health monitoring, one of the greatest challenges encountered when investigating correlations (whether by multiple regression or other means) is the selection of the set of independent variables that will comprise the regression model. Typically, we start by listing all known and hypothesized factors that might determine the value of the response variable ($Y$).[119] The result is often a lengthy list of variables (e.g., current speed, stream size, maximum annual temperature, nutrient availability, substrate, flow regime, channel stability, catchment area, etc.) that must be screened for their potential suitability for inclusion in the model. Any immeasurable variables (i.e., those too costly to collect or difficult to define, such as "flow regime" in the example above) are eliminated from the list, as are cases of covariation that occur between variables like catchment area and stream size.[119] Particular attention must be paid to the elimination of covariance in the model. Failure to do so commonly results in misinterpretation of the regression model because significance testing to evaluate the deterministic importance of each independent variable is meaningless when covariance exists.[41,68] Because we seek the most parsimonious model to describe the $X_1/Y$ relationship, variables in addition to $X_1$ are considered extraneous and are only included when they significantly clarify the relationship. Logically then, the next step involves narrowing the list of potential independent variables according to predictive power. Any variables that account for only a minimal amount of the variation between sites or times are discarded. In Equation 3.4, the magnitude of the various $\beta$ coefficients represents the relative predictive power associated with each of the independent variables included in the model. Tests of the strength of the correlation (the proportion of variance in the dependent variable explained collectively by all of the independent variables) and tests for lack of fit should follow.[68]

Although multiple regression alone cannot establish causation, it can expose a subset of environmental variables that are most responsible for the variability in a chosen measure of watershed health. In a pragmatic sense, when true controls and an appropriate level of replication are not feasible (because of costs, human resources, or time constraints) or are impossible (as is frequently the case in field bioassessments), multiple regression is one of relatively few tools for exploring causal linkages.[41]

Applied stepwise, multiple regression[68] can yield a subset of independent environmental variables that together can provide significant power for predicting biological (or other) measures. A detailed discussion of multiple regression can be found in Green's[68] text on biostatistics. A valuable guide and associated software package (Microsoft Excel® add-in) is also available on the Internet.[121]

## 3.4 Data quality assurance and control

Data quality assurance (QA) and quality control (QC) embody a set of checks and balances that ensure the following:

- Data generated by the study represent ambient conditions and answer the questions at hand
- Reliability of the data is independent of the collector
- Results are comparable to those from other assessments using similar methods

A data QA plan should be developed as a component of any watershed health monitoring program to set objectives for precision, accuracy, reproducibility, completeness, and comparability[19,122] of assessment information. Indeed, within the United States, government funding of volunteer monitoring programs is generally conditional on proponents implementing such plans.

QA and QC really begin long before any sampling is undertaken. Exploratory analysis, which provides an understanding of the variability of repeated measures,[41,49] is actually the first step because it provides a basis for sound design of experiments and surveys. In any assessment, the fundamental uncertainty that practitioners face relates to our ability to reliably characterize atypical variation. This conundrum is best expressed as the following question: If individual measures (e.g., abundance estimates, mean pollutant concentrations, biotic indices, etc.) exhibit variance over space, time, or both and are found to differ at two sites, do they differ because the value of the measure (and here we will assume also the health of the measure) is actually different at the two sites, or do they differ solely because any two samples would be expected to differ by as much even if collected at the same location at the same time?[41] Exploratory studies conducted before study design can characterize natural patterns of variability and help overcome problems such as determining how many replicate samples are required to allow reliable differentiation between sites or times. We direct practitioners to consult their favorite statistical text for direction on relevant exploratory studies. The following sections will deal primarily with QA and

QC techniques that are implemented during and after bioassessments and water chemistry sampling activities.

All data are variable. Just as natural systems vary according to taxa presence, altitude, temperature, and a myriad of other factors, variability also occurs within and among samples of environmental attributes. Use of QA and QC is not about eliminating variation. It is about ensuring that the data generated during a study are sufficient to reliably characterize watershed health. Apart from the difficulties associated with study design, which are discussed in some detail in Sections 3.3.1 through 3.3.4, there are many approaches that can be built into assessment programs through training of staff and volunteers and by implementing QA checks during data collection, analytical, and reporting phases of studies.

In bioassessment studies, one of the most significant QA problem areas is the area of taxonomic determinations, particularly for benthos and periphyton. The fact that many practitioners make identification errors, particularly at higher taxonomic resolutions (genus and species levels) is one of the reasons that *rapid bioassessment* protocols (which typically rely on coarser family- or even order-level determinations) gained much support in recent times. The best ways to prevent identification errors from occurring are through training and certification of practitioners. When using protocols requiring fairly rigorous (genus or species level) determinations, a reference collection of specimens checked by a taxonomic expert is a necessity. Laboratory protocols should require verification of voucher specimens with a taxonomic expert for any groups or taxonomic levels outside of a practitioner's level of training and certification.

Equally problematic among bioassessment studies and water chemistry and habitat surveys are difficult-to-detect transcription errors that may occur during analytical and archiving phases of studies. Some relatively simple QC protocols for addressing transcription errors include data sheets and database forms designed in an intuitive fashion and with regard to the order that measurements are made in the field. Within database applications, the use of *validation rules* and *lookup tables* in error-prone database fields is also simple and effective. Validation rules disallow entry of data that are not formatted correctly or are out of the acceptable range, whereas lookup tables allow data entry workers to select from a list of acceptable values, thus preventing typographic errors. We have found the latter to be essential for preventing problems later on when databases requiring entry of taxonomic names are queried. Similar error-fighting techniques can be employed in spreadsheet programs. Some of these are extremely simple, such as the use of split panes that enable column headers to remain visible regardless of where users scroll on the worksheet and that prevent data from being entered into incorrect locations. Other techniques (such as protecting cells containing standard formulae) are warranted when practitioners are not entirely familiar with the program they are using. Although redundant from a data-processing point of view, cross-checks on total counts (e.g., total number of data points, total richness, total number of sites) are useful for verifying that counts of individual species or sites have been entered correctly.

One key component of the QA plan for biological (and other) types of data includes development of watershed-based, regional, or even national databases. Complete with analytical modules and routines to assist practitioners in defining reference condition, these databases do the following:

- Improve the precision of biological assessments by assisting in the selection of reference sites with similar community potential[22] as proposed test sites
- Reduce the occurrence of data entry, transcription, and computational errors

For measures of physical habitat or biological attributes for which there is no known standard, the questions of accuracy and repeatability are troublesome and should be addressed through exploratory sampling before study design. One technique for ensuring sufficient confidence in assessment measures is to use sequential sampling techniques. Here, a larger number of replicate field collections is made than would be deemed necessary by exploratory studies. Samples are sequentially analyzed, and metrics are calculated until the desired level of precision is achieved (i.e., confidence intervals for the chosen estimators are of acceptable width). This technique is useful in assessments in which the costs associated with sample processing far outweigh costs associated with sample collection.

Regardless of data type, there are a number of techniques available to check for errors among archived data sets. In the case of biological data, the frequency of occurrence of a given species in a set of replicates from various sites can be cross-checked against specified acceptable limits based on exploratory work. Calculation of cumulative numbers for the different species collected can be compared with total numbers recorded for each replicate as a means of screening for potential transcription or summation errors. Within sites, ratios of the abundances of selected species to total numbers sampled can shed light on which counts or identifications are erroneous.

Checks for outliers are also commonly employed to screen for transcription, calculation, and analytical errors. Bivariate checks are extremely useful when one is screening for outliers. Scatter plots of variables against each other, over time or distance from a known point of impact, provide useful observations on whether the data follow logical patterns.[41]

Suspect values can similarly become apparent through regression with related variables. This technique is particularly useful when collections occurred over an obvious environmental, spatial, or temporal gradient. For example, abundance of a given taxa group may be related linearly to distance from a pollution source. If the source is relatively consistent over the sampling period, major departures (perhaps three or more standard deviations) from the regression line will highlight procedural or real environmental problems.[41] We echo Norris and Georges'[41] caution here that although screening for outliers that may represent erroneous data is a necessary process, outliers should only be deleted from data sets when "an *a priori* explanation of the aberrance is possible (e.g., misreading an instrument), when detailed scrutiny of equipment or records reveals the source of error or when the value is impossible." Furthermore, in highly skewed data sets, screening based on several standard deviations from the mean may result in a large number of suspect results. Calculating percentiles and scrutinizing values lying above (and below, in a two-tailed scenario) the 95th or 99th percentile may be a better choice in such situations.[41]

One of greatest sources of uncertainty in water chemistry data concerns the reliability of laboratory analyses. Fortunately, there are a variety of simple methods for assessing the precision of analytical results. Perhaps the most obvious are measures of central tendency (e.g., variance and standard deviation) that can be calculated for replicate samples having identical expected values, that is, that are collected very close together in time and space.[122] Most practitioners are familiar with measuring accuracy as the difference between the known concentration in a standard solution and the analytical result when the standard is analyzed for the parameter in question. A number of similar techniques include the use of spiked samples (samples with a known amount of analyte added), field blanks (analyte-free samples treated in the same manner as regular samples), equipment blanks (analyte-free samples used to check for analyte residues left in lab equipment from previous samples), and split samples (homogenized and subsampled fractions of a single replicate). All of the above techniques are useful for assessing accuracy of a protocol or of equipment or personnel. Although they are most familiar to those involved in water chemistry studies, one can imagine parallel or equivalent tests that could prove useful in studies using other indicators.

Detection limit and measurable range of both equipment and procedures are other topics of importance to data quality assurance. This is particularly true for water chemistry approaches, in which inferences hinge on the prevalence of specific pollutants. Calibrate and maintain instruments according to manufacturer's specifications!

## 3.5 GIS and modeling tools

GIS and modeling tools are important to both surveillance and performance evaluation. GIS is commonly used as a data management tool, a hypothesis generator (e.g., for generating hypotheses regarding interactions between land use and status of natural resources), and a reporting tool (for synthesizing complex data sets and generating maps and other presentation tools, etc.). Modeling is also commonly used as a hypothesis generator and predictive tool. Iterations of surface water phosphorus models, for example, can be used to predict impacts associated with land use change (most commonly development or implementation of some best management practice) or biotic responses to rehabilitation works. Figure 3.4 gives an overview of several hydrologic and water quality models used in North America.

| | | Model Name | | | | | |
|---|---|---|---|---|---|---|---|
| | | AGNPS | BASINS/ HSPF | GAWSER | STORM/ SWMM | ISWMS | QUALHYMO |
| Background | Sponsor | U.S. Department of Agriculture | U.S. Environmental Protection Agency | University of Guelph, Grand River Conservation Authority and Ontario Ministry of Natural Resources | U.S. Environmental Protection Agency | Greenland International Consulting Inc. and Nottawasaga Valley Conservation Authority | Technical University of Nova Scotia |
| | References | 140–142 | 143, 144 | 145, 148 | 147–149 | 150 | 151 |
| | General Type | conceptual distributed loading | conceptual lumped watershed | conceptual distributed watershed | conceptual lumped watershed | conceptual distributed watershed | conceptual lumped watershed |
| | Time Scale | continuous | continuous | continuous | continuous | continuous | continuous |
| | Spatial Scale | site to watershed | site to watershed | site to watershed | watershed | site to watershed | watershed |
| **Watershed Management Applications** | **Hydrology and Water Quality** | | | | | | |
| Rural Hydrology | Ground Water | | X | X | X | X | X |
| | Interflow | | X | X | | | |
| | Surface Runoff | X | X | X | X | X | X |
| | Snow Accum./melt | | X | X | | X | X |
| | Evapotranspiration | X | X | X | X | X | X |
| Rural Water Quality | Erosion/Transport | X | X | X | X | X | |
| | Soil Processes | | X | | | | |
| | Nutrients | X | X | | X | X | |
| | Bacteria | | X | | | X | X |
| | Pesticides | | X | | | | |
| | Other | | X | X | X | X | |
| | BMP Assessment | X | X | X | X | X | X |
| Urban Hydrology | Groundwater | | X | X | X | X | X |
| | Interflow | | X | X | | | |
| | Surface Runoff | | X | X | X | X | X |
| | Snow Accum./melt | | X | X | | X | X |
| | Evapotranspiration | | X | X | X | X | X |
| Urban Water Quality | Solids Transport | | X | X | X | X | X |
| | Nutrients | | X | | X | X | X |
| | Bacteria | | X | | | X | X |
| | Trace Organics | | X | | | | |
| | Other | | X | X | X | X | X |
| | BMP Assessment | | X | X | X | X | X |
| | **In-stream Process Modeling** | | | | | | |
| | Water Body Types Supported | | River, Reservoir | River, Reservoir | River, Reservoir | River, Reservoir | River, Reservoir |
| | Streamflow | | X | X | X | X | X |
| | Hydraulics | | X | X | X | X | X |
| | Point Sources | | X | X | X | X | |
| Solids Routing | Setting | | X | X | X | X | X |
| | Scour | | X | X | | X | X |
| | Particle Sizes | | X | X | | X | X |
| | Temperature/Heat | | X | X | | | |
| | Sediment–water Interactions | | X | | | | |
| Water Quality | Nutrients | | X | | X | X | X |
| | Bacteria | | X | | | | |
| | BOD-DO | | X | | | | |
| | Toxics/Metals | | X | | | | |
| | Other | | X | X | X | X | |
| | Plants | | Phytoplankton | | | | |
| | Reactions | | X | | | X | X |

***Figure 3.4*** Selected watershed management models used in North America.

Today's GIS software assists in study design, analysis, interpretation, and presentation of results. Computerized GIS models are of two types, *vector* and *raster*. In vector systems, the data are referenced according to ordered $(x,y)$ coordinates. Features in these models are represented as points, lines, or polygons (closed shapes formed by series of lines). Raster models store data in rows of uniform cells, with size and orientation of individual cells defined according to the feature's attributes. Raster models are generally considered preferable for manipulating classified image data but often contain less detail than vector data.[123] Inputting map data into GIS can be done

***Table 3.3*** Partial Listing of Currently Available GIS Software Packages[123,124]

| Product | Description | Vendor | Information Web site |
|---|---|---|---|
| ARC/INFO® | Vector based; full function | ESRI Inc. | www.esri.com |
| Microstation GIS Environment® (MGE) | Vector based; full function | Intergraph Corporation | www.intergraph.com |
| AutoCAD Map® | Vector based; full function | Autodesk Inc. | www.autodesk.com |
| MAPInfo® | Vector based; full function | MapInfo Inc. | www.mapinfo.com |
| MapFactory®; MFworks® | Vector based; full function | ThinkSpace Inc. | www.thinkspace.com |
| ERDAS® | Raster based; full function | ERDAS Inc. | www.erdas.com |
| ENVI® | Raster based; full function | Research Systems Inc. | www.rsinc.com |
| SPANS® | Raster/vector integration; full function | PCI Geomatics | www.pcigeomatics.com |
| IDRISI® | Raster based; full function | Public domain software suite | www.clarklabs.org/03prod/32frame.htm |
| GRASS® | Raster/vector; multifunction | Public domain software suite | www.baylor.edu/~grass/ |
| MapObjects® | Internet GIS software suite | ESRI Inc. | www.esri.com |
| GeoMedia® | Internet GIS software suite | Intergraph Corporation | www.intergraph.com/geomedia/ |

by digitization (either hand tracing or scanning) and can be a costly and time-consuming endeavor. The attributes of map objects (e.g., streams, roads, buildings, landforms, vegetative community types and sizes, etc.) are most often imported in database form. Data manipulation within a GIS can be done as soon as the spatial data and associated attribute data have been checked for accuracy and linked together. In most applications where data over a given area exhibit considerable heterogeneity, users define a classification scheme that sorts discrete areas into subsets of like data. Analysis procedures within GIS include the following:[123]

- Queries and multivariate models built from Boolean operators
- Determinations and sorting of areas with similar attributes, for example, similar soil types, land use, population density, etc.
- Assessment of spatial relationships between subsets of classified data to answer questions of adjacency (e.g., what subsets of data tend to be closely associated?), containment (what attributes are contained by or expressed in a given area?), and proximity (e.g., how close are objects to one another?)
- Network analysis, in which vector analysis of material or process routings enables practitioners to track processes of various types (e.g., flow of water or pollutants)

GIS systems can produce graphics to synthesize complicated geospatial data into meaningful presentations. Common outputs include sophisticated mapping schemes with color coding (or other mechanisms) to define common attributes and simulations over time or space to illustrate hypothesized impacts of human activities. Both can be powerful reporting tools.[123] Developing a successful GIS depends on *a priori* definition of the analyses required by the watershed management and monitoring teams because this will influence the choice of hardware and software, database requirements, and staffing and time commitments. There are numerous GIS software packages available (as indicated in Table 3.3), and many supplemental sources of information on GIS exist.[123–132]

# *Community education and awareness*

*Increasingly, provincial and municipal agencies are recognizing that public education and participation in the development of watershed management plans and implementation strategies are the key determinant of the success of these undertakings. If a rule exists, it is this: a public education and participation strategy should be developed early as an integral part of the watershed planning process.*

**—Ontario Ministry of Environment and Energy**[152]

## 4.1 Rationale

As Chapter 1 asserts, successful long-term implementation of a watershed health monitoring program is not achieved solely through good science. A well-rounded, closed-loop approach also incorporates an education and awareness program. In this chapter, we explain why these are essential and describe some techniques for maximizing the educational and community outreach opportunities associated with monitoring programs.

Here, we define *education* as a process whereby information exchange is used as a means of inducing a behavioral response. An excellent example of the types of behavioral shifts that we strive for as watershed managers is given in Chapter 5, which deals with attracting participants into partner networks that are ultimately responsible for providing means of cost recovery for a sustainable monitoring program. Community education, awareness, and participation are central to watershed health monitoring because they are building blocks for a bottom-up monitoring and management process that is capable of sustaining itself through grassroots support even in unstable political climes. Obviously, watershed communities cannot buy into a watershed health strategy unless they understand it and the issues that it addresses. Only educated watershed residents and stakeholders can contribute in a constructive way within a management process that is often clouded by misinformation, rhetoric, and politics. Educational messages and heightened awareness provide a vehicle for inducing behavioral shifts and other types of community action that are often required to resolve major watershed health issues identified through monitoring. Opportunities to participate in management activities (especially monitoring) solicit buy-in from the community, something that is critical if programs are to be sustained over the long term. Heightened awareness is also linked closely with the securement of the other aspects of our Closed-Loop Model. For example, political support escalates as politicians understand the financial, cultural, and other benefits associated with both monitoring the health of the watershed and using this information to maintain the delicate balance between human exploitation of resources and the ability to sustain ecosystem function. With general public support and backing by political leaders, significant hurdles to the establishment of a long-term funding model also become manageable.

Early in the watershed management and monitoring process (i.e., during the first round of the watershed planning process), information is typically related to surveillance monitoring, for example, as follows:

- Summarization of spatial and temporal trends in watershed health and identification of constraints and potential management strategies for maintaining balance
- Delineation of environmentally significant and unique features, ecological functions of the area, and how they interact as a system with elements of the human side of the watershed health equation such as economics, opportunities for cultural expression and recreation, and population growth

Later on in the watershed management process, once the watershed plan has been implemented, more of a balance is required in relation to the delivery of surveillance and performance evaluation components of the monitoring program. Performance information is required to let all parties know how programs are doing in relation to planning targets so that adaptation is possible and to maintain the participation of partners. Once management activities are well under way, information is also required by watershed residents, businesses, and other partners regarding what actions they can take to assist in the maintenance of a sustainable condition.

## 4.2 Techniques for educating the watershed community

In this section, we detail some of the many possible options for educating the watershed community. Each approach relies on the following assumptions:

- Key target audiences (which usually include watershed residents, public and private sector stakeholders and partners, politicians, and the lead watershed agency) are identified.
- Current information appropriate for each distinct audience is delivered on an ongoing, consistent basis.

To ensure that the intended educational response is achieved, information should be presented on a factual, issue basis. Constraints on watershed health and potential conflicting uses need to be identified based on science. The forum for information exchange must be solution oriented, and any interjections based on rhetoric or misinformation must be dealt with expediently.

In Sections 4.2.1 to 4.2.7, we offer several techniques for delivering educational information to the general public.

### 4.2.1 Community-based monitoring

Perhaps the most obvious means of delivering an educational message to the general public is to provide the opportunity for direct involvement in management activities. Monitoring programs (especially watershed-wide surveillance monitoring aspects) offer an excellent opportunity for education and developing a sense of ownership among residents. Refer to Chapter 5 for a more detailed treatment of the role of participatory opportunities within the closed-loop approach.

### 4.2.2 Seminars and workshops

Formal, strategic gatherings are excellent venues for educating and enhancing watershed awareness among audiences that frequently include politicians, the watershed community (i.e., the general public), partners, and the management team. We have used seminars and workshops effectively for target setting, issue identification, performance evaluation, and policy development related to all aspects of watershed health monitoring. Such organized events can also provide delegates with opportunities to discuss common issues and share solutions and expertise across watershed boundaries.

When choosing venues for organized events, keep in mind the expectations of the various audiences and strive for a central location. Local watershed businesses (hotels, resorts, conference centers, etc.) or government buildings are excellent choices because mutually beneficial partnering opportunities may arise at any time!

Slide shows can be very effective and interesting means of delivering watershed monitoring messages in less formal settings and can subsequently be adapted to printed media, posted on Web sites, or used as handouts and flyers.

### 4.2.3 Field trips

Field trips can be an inexpensive and effective means of educating small- to medium-sized groups and are particularly useful for training participants on specific monitoring activities. A checklist of considerations to keep in mind while planning field trips is given in Figure 4.1.

Timing is critical for strategic field trips. This is particularly true when intended audiences include politicians (who typically agonize over prioritizing opportunities for appearances at a multitude of community events). Consider, for example, a strategic tour of monitoring sites to be held at budget time and aimed at illustrating surveillance and performance measures along the route (i.e., explaining why they cost so much!). To secure attendance at such an event, it may pay to ensure a media presence and

- ☐ Appropriate date and suitable alternative (rain date)
- ☐ Key representatives from all intended audiences invited
- ☐ Attendance confirmed well in advance of departure date
- ☐ Appropriate transportation secured
- ☐ Briefing notes/curriculum (highlighting key messages) with indication of how to obtain more information or get involved
- ☐ Arrangements made for appropriate technical personnel to meet on-site for demonstrations
- ☐ Refreshments ordered and comfort stops scheduled
- ☐ Press representatives invited and encouraged to cover the event through video or print media
- ☐ Appropriate tour guide secured (e.g., chair/manager or professional facilitator)
- ☐ Personnel secured to document proceedings and followup with reports to partners, politicians, the watershed management team, and the press
- ☐ Tokens of appreciation/memorabilia (t-shirts? buttons?) purchased

***Figure 4.1*** Checklist of considerations for planning educational field trips.

photo opportunities for supportive officials—another example of linkages among political support, education, and cost recovery.

Technical monitoring teams may also benefit from field trips, particularly from the point of view of fostering a better understanding of the linkages between environmental and sociocultural watershed health indicators (remember our definition of watershed health given in Figure 1.1) and investigating opportunities for integration of monitoring techniques. Program participants themselves are frequently targeted as audiences for field visits to monitoring sites, the purpose of which may be to foster a better understanding of how cash and in-kind resources generated through a partnering model (as discussed in Chapter 5) are allocated through the monitoring program. Where monitoring is performance related, tours frequently include destinations that are useful illustrations of watershed management projects (often restoration works or examples of implemented best management practices). These events strategically promote buy-in and reinforce commitment. As we discuss in Chapter 5, partners contribute because they want to make a difference and because they can extract many tangible benefits from their participation. Seeing their contributions in action goes a long way to ensuring continued support.

### 4.2.4 Media coverage, marketing, and promotion

Many practitioners recognize the powerful educational potential of direct reporting through press coverage, marketing, and promotional messages. Such reporting can take many forms, including community newsletters; television content (e.g., televised news, call-in shows, talk shows); radio interviews and documentaries; newspaper stories, advertisements, and inserted flyers; promotional videos, events, and fundraisers; and Internet sites. Methods for boosting uptake and gaining coverage are equally abundant. These include engaging in direct fee-for-service arrangements (as in newspaper or television advertisements); issuing media releases; holding special events (such as tours and seminars); scheduling the appearance of a political figure; requesting interviews with members of the press; linking to related Web sites; and flyers and signage.

Of course, practitioners must be familiar with which media types best reach various audiences. Politicians are often motivated to respond to information that either supports or detracts from their position on key watershed issues. Consider televised or printed media coverage of events with political implications or with political attendance. Challenges between politicians can generate much attention, for example, one county representative challenging another to adopt some philosophy into an official plan or budgeting exercise. The general watershed community is often best reached through traditional press coverage, targeted mail-outs, personalized invitations, television coverage, and tax note insertions. Potential partners can often be reached by highlighting ways that they can benefit from support for or involvement in particular aspects of the program (often stressed are future growth or marketing opportunities, enhanced public image, and personal satisfaction).

Printed materials are often one of the first educational tools contemplated by practitioners. Booklets and flyers offer straightforward and concise information to the public and can represent a tremendous educational opportunity. Great attention to design is required, however, to ensure that items are read in today's world of information overload. A concise style is also important because of the costs for materials and distribution. Reports and pamphlets are useful for delivering a concise message to a large number of people via regular or electronic mail but may have minimal ability to generate a significant behavioral response unless they are closely followed by messages in alternative forms. Another form of printed media includes informative signage that can be used in recreational and other community areas. This medium offers little content but may have value associated with its ability to provide concise messages to a relatively wide audience.

### 4.2.5 School curriculum

Schools represent the best means of delivering information to future decision makers, and (as experience with recycling programs in Ontario illustrates) young people can be powerful catalysts for broad changes in behavior—the types of changes that are frequently required to tackle major threats to the watershed health balance. With community support and participation in the monitoring program, it can be relatively easy to incorporate principles of watershed management into school curricula. The process typically entails enlisting a group of teachers to write segments or units that fit into the educational curriculum framework for the appropriate levels of students to be involved. We advise that teachers do this through close communication with representatives of the various groups involved in the monitoring program (although traditional texts on watershed monitoring may have suggested involving *technical or scientific* monitoring personnel in the curriculum-writing process, we believe that personnel responsible for managing the other community, cost recovery, and political aspects of the program are equally important and contribute to a multidisciplinary curriculum). Subsequent stages include broad peer review by teachers, followed by proposal of the curriculum packages to the appropriate school administrators.

Less formal activities for utilizing schools to disseminate monitoring program messages include guest lecturing programs, field trips, and job-shadowing or cooperative-education placements for students.

### 4.2.6 Utilization of community organizations and service groups

Collectively, public interest and not-for-profit groups within the watershed often reach out to a large percentage of the community. They represent an audience that can be targeted through many of the techniques discussed in this section and are particularly effective at disseminating educational messages in rural or ethnically diverse areas.

An excellent means of boosting information transfer through such groups is to strive for maximal participation through a partnering cost-recovery model for the program. As discussed in Chapter 5, such relationships help ensure long-term sustainability of the program and also result in information flow back to the membership of contributing groups.

### 4.2.7 Computers and the Internet

High-tech alternatives exist for message delivery. Compact discs (CD-ROMs) may be used with computers for communication and educational purposes and are a relatively inexpensive way to disseminate large amounts of information. They are therefore ideal for annual reports or other documents containing information that is relevant in the context of adaptive management (where adaptation is obviously contingent on achieving behavioral change). Web sites are powerful tools because they facilitate widespread access to information by a computer-literate audience. Indeed, most watershed management groups in the United States and Canada host them. Typical program Web sites have background sections that describe the watershed community; physiography of the basin; goals and objectives of the program; and the governance model and funding structure for the watershed management program. Other common sections include monitoring summaries that showcase the status of watershed health; trends, hot spots, or areas of concern identified through the monitoring program; instructions on how to get involved or share opinions on watershed issues; and descriptive maps and other graphics. Individual Web sites should be linked to sites hosted by other contributing partner groups to facilitate navigation to pertinent information and to improve the priority scores assigned to the site through Internet search engine keyword searches. Most important, Web sites must be kept current. The modern Web surfer has little patience for out-of-date information.

A watershed geographic information system (see Section 3.5) is another high-tech tool that should be considered as a valuable data analysis and reporting tool. These information management systems can support the delivery of watershed health messages through various media, especially thematic mapping.

## 4.3 Avoiding pitfalls

Practitioners are exposed to numerous pitfalls while developing and implementing communication and educational strategies. In this section, we describe some guidelines for avoiding typical setbacks.

Regardless of the communications tool being used, practitioners must understand their audience. In the case of a farming community, for example, where traditional modes of communication are the norm and few have access to computers, do not rely on e-mail circulation of newsletters! Recognize opportunities and be creative. In the same hypothetical farming community, whereas lack of computer literacy may be viewed as a communication barrier, close relationships between neighbors may result in extremely fast transfer of information via word of mouth. So a wiser approach may be to attend fall fairs and simply talk with residents.

Ensure that issues raised by the public are dealt with seriously by the monitoring (and overall watershed management) team. If residents believe that their interests are being defended, they are more likely to contribute to and take ownership of the monitoring program. Persistent pleas for assistance without appropriate delivery will quickly lead to requests being ignored. When communicating with watershed residents, be mindful that people expect returns on their investments. Be sure to express to them what they stand to gain from participation in monitoring and other management activities (as we stress in Chapter 5) or for changing the way they conduct their daily business if that is what educational messages strive to achieve.

The old adage, "you never get a second chance to make a first impression," speaks to the need to draw your audience into a compelling discussion early on in a process. As you work with your audience, establish trust and credibility by listening. Be honest, candid, and speak clearly and with compassion. Maintain consistency and equality in the monitoring and communications program to build trust and ensure that the community is united in the watershed cause. We advise hiring a full-time liaison person who is responsible for coordinating an education, information, and awareness program (often this is through the lead watershed agency) whose job it is to uphold these principles.

A common concern raised by public participants is that instances of noncompliance negate good work done elsewhere. A critical tool to establishing credibility is providing contact information for residents to report concerns or infractions.[155]

New real estate purchasers represent an important audience for watershed health information because they are often new to an area and interested in learning more about local issues while they become integrated into the community. They also need to know what their own roles and responsibilities are in terms of local management programs. We suggest that communication and educational efforts be targeted to all home buyers for these reasons.

An important way to increase public understanding of watershed issues is to incorporate the message that sustainable environment and economy are inseparably linked. Environmental restoration can be a major economic development tool. Environmental restoration can serve as a new industry within the watershed, providing training and skilled jobs.

Recognize that a favorable rapport with the media is a powerful educational asset and that considerable effort may need to be expended to ensure that reporters provide a balanced treatment of topics that is not tainted by rhetoric or sensationalism. We have seen weekly columns reach wide audiences and contribute greatly to general awareness of monitoring and other management activities. Their focus should be on success stories; coverage of citizens and community groups participating in projects; new planning and development practices to protect and enhance the environment (technology transfer); discussions of pertinent legislation and local watershed health issues; and analyses of local government decisions that have implications for implementation.

# Cost recovery through partnership

*For the monitoring and restoration of the Great Lakes, useful strategies have actually been developed in the past, but their implementation has been poor because of faltering commitment, cutbacks in funding and weak project management. Ministries have on some issues misused progress reports as public relations exercises, focusing on their achievements. They have failed to articulate and analyze roadblocks to achieving their targets and as a consequence have not been able to practice adaptive management.*

**—Environmental Commissioner of Ontario**[153]

## 5.1 Rationale

In Chapter 1, we discussed the fact that monitoring activities must be part of a cyclical watershed management process to enable the adaptation of management programs as watersheds change due to growth, political pressures, and changing issues. We argued in Chapter 3 that there must be long-term commitment by the watershed management group to provide key scientific deliverables associated with surveillance and performance evaluation. Because of the cyclical, ongoing nature of watershed management and the ever-present need for adaptability, it follows that a successful watershed health monitoring program is contingent on the development of a sustainable cost recovery model (hence its inclusion in the Closed-Loop Model, as illustrated in Figure 1.3).

In an ideal situation, funding would arise year after year from a single, dedicated source. From our experiences in North America, however, this scenario seldom pans out. Commonly, groups look to government sources or private sponsors (such as philanthropic groups) to provide funding for watershed monitoring activities. As discussed in Chapter 2, the problem with this approach is that the benefactor's perception of the importance of the monitoring program (hence worthiness of continued support), as well as the funder's liquidity (i.e., ability to pay), is dependent on a range of economic and political factors that are entirely beyond the control of the group receiving the support. Planning for growth and development of monitoring programs then is extremely risky under such a funding model. For monitoring and other watershed management activities, we advocate an approach whereby cost recovery is achieved through the development of partnerships. Partnering efforts overcome many of the shortfalls associated with programs funded by a single entity and represent opportunities to concurrently develop other key components of the Closed-Loop Model: political linkages; a forum for public education and involvement; and a more secure future for surveillance and performance-monitoring activities. One example of a partnering model that has been successfully applied in Ontario is described below.

## 5.2 A cost recovery and partnering model

In this section, we explain the components of one cost recovery and partnering model (Figure 5.1) and discuss the interrelatedness between cost recovery (through partnership) and the other aspects of the closed-loop approach.

In the cost recovery model proposed in Figure 5.1, stakeholders may be categorized as either advisory partners or resource partners. Advisory partners are any stakeholders or subgroups of the lead watershed agency that are capable of providing support in the form of strategic direction or assistance with implementation of the monitoring program. Resource partners are those parties (usually stakeholders) capable of contributing cash or labor

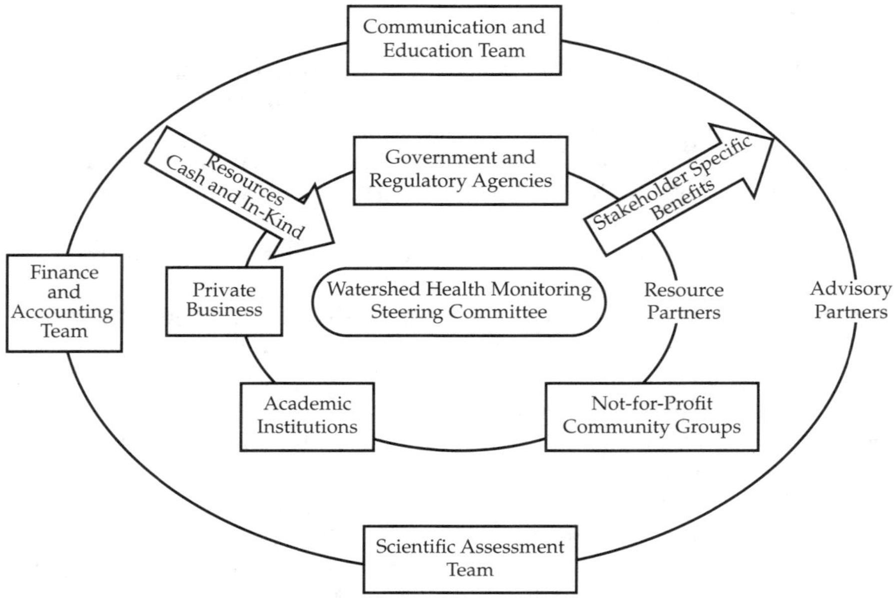


***Figure 5.1*** A cost recovery partnering model that provides a framework for stakeholders in watershed management to provide cash and in-kind resources to one group (often the lead watershed agency or monitoring steering committee) while extracting tangible benefits in other "currencies," such as information, influence, or exposure in the community or credibility. (Adapted from models developed by Sustainable Development & Monitoring Inc., Waterloo, Ontario, Canada.)

to directly support monitoring activities. The hub of the model is a steering committee that oversees all aspects of the watershed health monitoring program and reports directly to the lead watershed agency. For purposes of facilitating partnership and consensus building, the steering committee generally comprises representatives from resource and advisory groups in the outer rings of the model.

Although the net flux of cash and in-kind resources is toward the center of the model, for allocation by the steering committee, it is paramount to understand that there is also a net flow of benefits outward from the center. These benefits are measured in whatever "currency" is valid according to each contributing partner. Commonly cited examples of such currencies include information, influence, credibility, and personal satisfaction. It is also critical to understand that the role of the lead group at the center of the model is more than just allocation of resources. The group must also maintain a balance between the net value of incoming resources vs. outgoing benefits to partners. This is true because imbalances lead to instability and the ultimate breakdown of the model. On one extreme, partners become resentful of receiving diminishing returns on their investment while their contributions exceed perceived benefits. Conversely, scientific assessment of the feedback loop to watershed managers (i.e., the program's capacity for adaptation) falters because of lack of inward-bound resources.

Although the model presented in Figure 5.1 suggests that the partner network itself represents a boundary for the type and amount of contributions that can be made to the program, we should clarify that part of the function of the stakeholder group (with its varied collection of expertise) should ideally function as a catalyst for innovation in funding and cost recovery scenarios. Some examples of innovative ideas proposed within such forums in Ontario include the following:

- Economic incentives or tax relief options for watershed community residents who contribute resources to the monitoring program
- Introduction of new taxes or development levies earmarked for watershed monitoring and management
- A shift in government-funded monitoring programs from capital budgets to more far-sighted operational budgets (see Case Study 1)
- Mechanisms for allocating appropriate value to watershed resources in meaningful economic terms

Models such as the one described in Figure 5.1 tend to be extremely dynamic and are dependent on an excellent communication infrastructure. There must be plasticity in the model to allow for changing partner roles and foresight by the lead group to manage for succession of partners and expansion of programs over time. Each partner must contribute toward the overall goal in its own way, as permitted by its own mandates. As long as the lead group can satisfy these requirements, long-term sustainability is possible. Indeed, we have seen this approach successfully applied in Ontario (as discussed in Case Studies 1 and 3) and commonly used by watershed management groups across the United States. Further, toward understanding the dynamics of the model, consider that the role of the lead watershed group is to manage the personal and financial connectivity between partners and to initiate a snowball effect in which multiple stakeholders amplify the flux of resources through communication. Members of the lead watershed group must have an understanding of what benefit currencies are relevant to each partner.

Partnerships are successful for a number of reasons. Individuals may contribute because their jobs involve such cooperation. Many people enjoy working with others and meeting new challenges. Partners may see the potential for professional and personal growth, as well as a sense of accomplishment. External factors can also motivate partnerships, including public expectations and organizational mandates for cooperation. Informal, social interaction can provide the glue that holds a partnership together. To give some simplified examples, academic institutional partners may value gained public support, exposure before potential research funding partners, and numerous opportunities for involvement of graduate students. Community (not-for-profit) groups may value enhanced credibility and opportunities for networking, whereas their members may derive personal satisfaction from contributing to the management of watershed resources. Government agencies may see an opportunity to magnify the deliverables associated with expenditures of core programs through the phenomenon of "matching dollars" while gaining information needed to shape policy and develop a rapport with taxpayers.

As alluded to in the opening paragraphs of this chapter, the strengths of a partnering model in relation to long-term sustainability of funding include the following:

- Less dependence on the political climate and affluence of a single large contributor
- More control and influence over the flow of cash and other resources because the lead agency is the hub of all partnering relationships
- Opportunities for concurrently bolstering other aspects of the Closed-Loop Model with activities that directly boost available resources (education and awareness is required to bring in new partners, and political support increases as partner groups represented increase, etc.)

Further information on the intricacies of managing a partnering cost recovery model is presented in the following section.

## 5.3 Guiding principles

As indicated above, it is important that partnering frameworks allow for changing roles of participants and recognize that partnerships are more than just an opportunity to secure funding. Indeed they represent a powerful forum for facilitating work on other fronts of the closed-loop approach. Table 5.1 illustrates the key contributions that individual partners can make and meaningful benefits that they can extract from the partnering process based on experiences in applying the model in Ontario.

Of course, considerable challenges arise (as illustrated in Table 5.2) when managing large groups of individuals (often with competing or contrasting interests) toward a common goal. Indeed, if group dynamics are not managed skillfully, partnerships may be unsuccessful for a variety of reasons, including the following:

- Pessimistic atmosphere arising from past failures
- Lack of commitment
- Worry about lost independence
- Lack of credit for own contributions
- Personality conflicts
- Power struggles or turf battles
- Partners that do not agree on realistic roles and responsibilities
- Differences in cultural and personal values

***Table 5.1*** Typical Stakeholders Involved in Cost Recovery Partnerships; Emphasis Is Placed on Discussion of the Common Contributions and Benefits Extracted by Each Group as Well as a Treatment of How Various Partners Fit into the Broader Application of a Closed-Loop Approach to Watershed Health Monitoring

| Partner group | Typical contribution | Typical rewards extracted |
|---|---|---|
| Media | Direct role in reporting the workings of the partner network and attracting new participants; contribute dramatically to the larger public awareness and education campaigns of the lead watershed agency | Access to long-term supply of "good stories" of interest to local audiences; enhanced credibility (important to many media representatives who struggle to balance human interest stories with coverage of events of human and environmental tragedy) |
| Watershed residents | Good local information, in-kind labor and cash contributions; resident participation is also valued by other resource management wings that seek to improve watershed health through remedial projects on private lands | A greater appreciation for and understanding of local environmental health; personal satisfaction gained through participation in initiatives affecting the local community; greater access to programs with implications for boosting property values (e.g., fisheries rehabilitation projects on riparian properties) |
| Chambers of Commerce and local businesses | Cash and expertise; certain businesses (i.e., the agri-food and development industries) may be directly involved in other watershed management programs as key stakeholders; this involvement is often key to whether long-term political support is maintained | Opportunities to network within the community and attract prospective clients; personal satisfaction related to contributions within the community and a greater public perception as a "good corporate citizen" |
| Not-for-profit groups | Cash and in-kind support for all aspects of monitoring, community outreach, and educational activities; these groups are often well connected to political and business leaders that are valued resources for other aspects of the closed-loop approach | Enhanced credibility within the community and opportunities to participate in projects of joint interest to the lead watershed group and their own membership |
| Elected officials | Frequently an advisory role; particularly important to the application of the closed-loop approach because of political connectivity | Enhanced credibility as "green" politicians and opportunities to network with representatives from a variety of sectors within the watershed |
| Government and regulatory agencies | Direct funding and advisory roles (cash and/or expertise) | Opportunities to work toward departmental mandates while magnifying deliverables through matching community resources |

***Table 5.2*** Common Destructive Forces within Group Settings That May Be Overcome by Skillful Management and Fostering of a Team Environment within a Stakeholder Network

| Counterproductive element | Management considerations to promote constructive interaction |
|---|---|
| Lack of time or other resources (many partners will have other commitments and availability may fluctuate temporally); low levels of commitment or interest (this can happen if the effort gets bogged down or partners are not kept active); reluctant partners (most groups have one or more members who seldom contribute; problems arise unless these partners are encouraged to be active in some way) | Manage for succession of partners within the network and strive for a measure of redundancy within the group to fulfill the role of key partners during particularly busy times; foster a team atmosphere and offer positive feedback to contributors |
| Individualism (to many, the idea of working together is contrary to beliefs in self-sufficiency and competition; some people tend to feel it is a sign of strength to be able to solve problems on their own); overbearing or dominating partners (some partners—often those with authority or expertise—have too much influence over a partnership and can discourage discussion or criticize others' ideas); conflicting participant agendas; rush for accomplishments (some partners may push to do something, either because they are impatient or are pressured from elsewhere; these partners often reach their own conclusions before the rest of the group has time to carefully consider all options); floundering (trouble starting and finishing projects); lack of flexibility | Ensure goals are clearly stated; foster a team atmosphere with decision making through consensus |
| Loss of autonomy or recognition (people often worry that partnership means a loss of freedom or control over personal priorities and activities; some also worry that they may not get enough credit for the work they do within a partnership | Strive to maintain balance between individual contributions to and extraction of benefits from the group; develop a formal recognition program to honor participants |
| Fingerpointing in times of crisis; feuds and competition between partners; rhetorical monologue and unquestioned acceptance of opinion as fact; attribution and criticism (assigning negative motives to partners with opposing, and often not well understood, views); digression and tangents (some digression may be useful in generating innovative new ideas, but often digression simply wastes time; unfocused discussions can result from poor leadership) | Maintain a team atmosphere and focus on solution-oriented dialogue |

It is important for the lead watershed group to develop a procedure for consensus building as a means of avoiding alienation of any stakeholders from the partner network while the group works toward achieving goals through a series of decisions. To this end, all major interests should be identified early on in any decision-making process, with subsequent negotiations centering around finding common ground and working toward consensus through education, compromise, and all other means. All partners should be encouraged to take responsibility for the actions of the group and should be involved in mutual information sharing as a precursor to innovation and adaptation.

Obstacles that arise through group dynamics can consistently be overcome through skillful management and wise intervention aimed at fostering a team atmosphere. When establishing a partnership network, it pays to discuss ground rules for participants (e.g., descriptions of what contributions are expected, codes of practice for members, definition of roles and responsibilities). When problems do arise, they should be addressed as a group challenge (rather than as an individual problem). This goes against common tendencies to blame individuals for problems because it recognizes that many problems occur through flawed human relationships. When group problems boil down to the incompatibility of one individual (we have seen this in a number of situations), we offer the following series of potential actions arranged in escalating severity and frequency of disturbance within the group:

- Do nothing (infrequent or minor disturbances only).
- Give constructive feedback to the disruptive participant and illustrate how compromises are narrowing the gap toward consensus.
- Talk informally with disruptive partners outside the group setting and discuss general concerns at the beginning of a meeting without pointing out particular partners.
- As a last resort (once other approaches have failed), the leader may need to confront the person in the presence of the group or table a motion with implications for the continued participation of the disruptive party.

Innovative partnerships that encourage participation from each sector of the watershed community require hard work and foresight to establish; however, invested time and energy pay significant dividends in the form of sustainably pooled ideas, expertise, and resources.

# Case Study 1

# The Laurel Creek Watershed Monitoring Program

The Laurel Creek Watershed Monitoring Program (LCWMP) is an excellent example of how application of the Closed-Loop Model can lead to remarkable results. The LCWMP has been developed and implemented successfully since 1996, with continuous municipal funding and resources and with incredible political support from the City of Waterloo Council. Over the years, the city council has also adopted several new watershed management policies to address issues identified through monitoring. The University of Waterloo and the local development industry have contributed significantly toward the creation of comprehensive and scientifically sound assessment protocols for each of the chosen watershed health indicators. Responsibility for the implementation of the community awareness and education component of the LCWMP has been threaded throughout the municipal departments in addition to the various partners. That the LCWMP continues to be implemented and the project team continues to strive toward its objectives indicates that the partnership approach for sustaining the program is working well. In fact, the program continuously improves through implementation of annual recommendations. The City of Waterloo is continuing to explore other creative cost recovery mechanisms.

In 1993, the Grand River Conservation Authority (GRCA, the local lead watershed agency), the City of Waterloo, and others completed the Laurel Creek Watershed Study (LCWS). The study was designed to identify terrestrial and water resource processes, the impacts of resource use, and the potential impacts of urban and rural land use changes on the natural environment. The study recommended an effective long-range watershed management strategy. The LCWS also recommended the implementation of a monitoring program to assess the carrying capacity of the watershed and provide information to decision makers to make sure that this capacity is not exceeded. The City of Waterloo embraced this proactive recommendation by developing and implementing the three-part monitoring program shown in Figure CS1.1.

Furthermore, the City of Waterloo accepted these recommendations through the inclusion of legally binding watershed/subwatershed policies in the Official Plan and subsequent regulatory planning and engineering documents (e.g., Plans of Subdivision, Site Plans, and District Plans). The monitoring program was designed to identify the condition of various environmental features (both aquatic and terrestrial) within the watershed and provides a basis for evaluating environmental health responses to future development (if any).

*Figure CS1.1* The City of Waterloo, Ontario's three-part monitoring system.

## CS1.1 Background

The Laurel Creek Watershed includes an area of approximately 74 km$^2$, of which 80% is within the City of Waterloo in Ontario, Canada. Existing land uses are placing pressure on the environmental resource features in the watershed. The mission statement for the LCWS was "to achieve sustainable development which is aimed at maximizing benefits to the natural and human environments on a watershed basis." The need for environmental monitoring in the Laurel Creek Watershed was identified in the LCWS as a critical means of ensuring that the carrying capacity of the watershed is not exceeded by approved development. Carrying capacity was defined as a level of development and landscape manipulation "which does not result in any further degradation of ecosystem health and which allows for extensive rehabilitation and enhancement." The LCWS recommended that the status of several components of the watershed system be monitored on a regular basis to evaluate whether the carrying capacity is respected. As a result, the LCWMP was developed by the City of Waterloo to assess the group of indicators illustrated in Figure CS1.2.

## CS1.2 Mission statement

The mission statement for the LCWMP was "to maintain and improve the health of the watershed by ensuring that the carrying capacity of the watershed is not being exceeded by approved development."

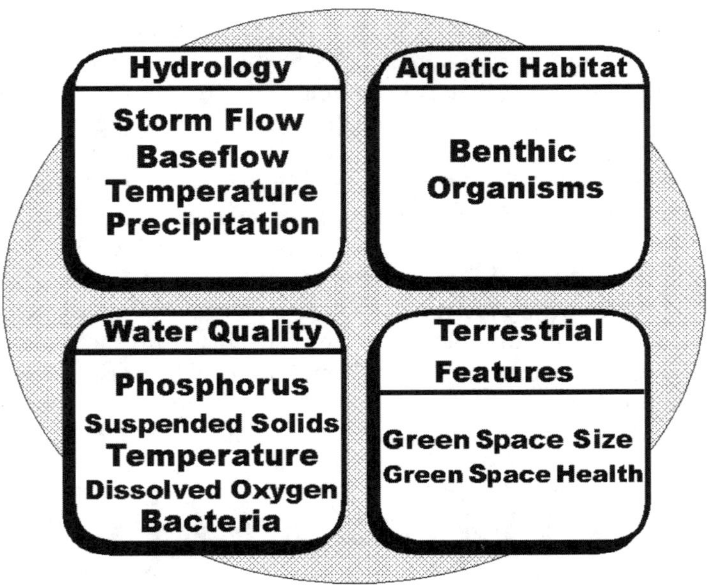


*Figure CS1.2* Indicators used in the LCWMP.

## *CS1.3 Program objectives*

The LCWMP objectives were as follows:

- To detect changes in the environmental health of the watershed
- To utilize monitoring data to maintain and improve the health of the Laurel Creek Watershed
- To determine the effects of land use practices occurring in the watershed
- To promote cooperation, communication, and partnerships with the development industry, citizens groups, public agencies, neighboring townships, and other relevant organizations
- To increase public awareness and encourage community stewardship
- To develop a comprehensive database for future comparison and analysis

Implementation of the LCWMP was achieved through shared responsibility of public agencies and developers, as illustrated in Table CS1.1.

The LCWS targets were chosen to minimize flood risk, avoid increased stream bank erosion, maintain adequate base flow levels, and prevent degradation of fisheries and stream and reservoir aesthetics, as illustrated in Table CS1.2.

***Table CS1.1*** Partner Responsibilities during Implementation of the Laurel Creek Watershed Monitoring Program

| Responsible parties | Purpose |
|---|---|
| **Part 1: System monitoring** | |
| Municipality and other public agencies having jurisdiction | Monitor the Laurel Creek Watershed ecosystem to establish long-term baseline condition; compare future ecosystem conditions with this baseline and identify trends |
| **Part 2: Development monitoring** | |
| Developer | Regularly monitor and maintain environmental conditions and facilities; provide mitigation measures where indicator targets and objectives are not met |
| **Part 3: Post-development monitoring** | |
| Municipality and other public agencies having jurisdiction | Regularly monitor and maintain environmental conditions and facilities; provide mitigation measures where indicator targets and objectives are not met |

***Table CS1.2*** Targets for New Development under the Laurel Creek Watershed Monitoring Program

| Indicator | Target |
|---|---|
| Flow discharge | Peak flows during and after development should match peak flows before development so that flooding does not result |
| Base flow | Maintain minimum predevelopment base flow levels to streams and wetlands such that aquatic habitat is maintained and water temperature requirements are met |
| Precipitation and air temperature | N/A (required for use in GAWSER model or Storm Water Management Model) |
| Water temperature | Warm-water fishery (smallmouth bass): maximum in-stream temperature of 26°C (June 1–August 1) and 29°C (August 1–October 31)<br>Cold-water fishery (brown trout), maximum in-stream temperature of 22°C (April 1–October 31) and 14°C (November 1–March 31) and minimum in-stream temperature of 4°C (November 1–March 31) |
| Phosphorus | Upstream of reservoir, maximum of 0.03 mg/l; downstream of reservoir, 0.05–0.08 mg/l |
| Suspended solids | Storm event mean concentrations of less than 25 mg/l |
| Dissolved oxygen | Warm-water fishery (smallmouth bass), minimum of 5 mg/l; cold-water fishery (brown trout), minimum of 6 mg/l |
| Bacteria | Upstream of reservoir, maximum of 100 CFU per 100 ml of *Escherichia coli*; downstream of reservoir, maximum of 200 colony forming units (CFU) per 100 ml of *E. coli* |
| Benthic organisms | Maintain populations of benthic invertebrates such as mayflies, stoneflies, and caddisflies that indicate relatively clean streams |
| Green space size | Maintain green space size and maintain corridors between woodlots and wetlands for migration of plant and animal species |
| Green space health (includes vegetative and wildlife health) | Maintain diversity of natural vegetation; improve vegetation quality and diversity along degraded stream corridors; maintain size, quality, and health of wildlife communities, including woodlot area–sensitive species (e.g., red shouldered hawk) and wetland area–sensitive species (e.g., bull frog, snapping turtle) |

Continued implementation of the program will enable the municipality to identify environmental degradation or improvement, to recommend and perform mitigation measures where necessary, and to establish the effectiveness of mitigative measures and best management practices.

## *CS1.4 Program development*

The following sections highlight key milestones in the development of the LCWMP.

### *CS1.4.1 Initial stages*

With the support of City Council and several legislative policies, an interagency team was established by the City of Waterloo to develop the program. Membership included representatives from the City of Waterloo, Region of Waterloo, Ontario Ministry of Natural Resources, Ontario Ministry of Environment and Energy, GRCA, the University of Waterloo, and private consultants. The team worked toward the resolution of the several key issues raised in the Westside Watershed Discussion Paper,[154] as shown in Table CS1.3.

### *CS1.4.2 Task team development and partnerships*

To efficiently devote resources to the technical, communications, and financial issues, the interagency team was partitioned into six subgroups known as "task teams." The six task teams included four ecosystem health indicator teams (in charge of hydrology, water quality, aquatic habitat, and terrestrial features), a data storage and analysis team, and an education and community awareness team. Through this subteam structure, each team addressed its own issues to work toward the synthesis of an overall monitoring program. Although resources were limited, the devel-

***Table CS1.3*** Key Issues Addressed during the Development of the Laurel Creek Watershed Monitoring Program

| Technical issues | Communication issues | Financial issues |
|---|---|---|
| Selection of ecosystem health indicators | Database maintenance responsibility | Funding sources and volunteer efforts |
| Establishment of targets | Method of reporting | Cost–benefit ratios for various technical and communications options |
| Identification of potential sources of variability and error | Mechanism to provide data to agencies and the public | |
| Selection of appropriate sampling devices and techniques | Videos, marketing | Partnerships |
| | Data storage | Stewardship |
| | | Budget |

opment of the LCWMP was completed in a timely manner, largely because of partnerships offering in-kind assistance, equipment, chemicals, and expertise.

## CS1.5 Pilot study

The following sections detail a pilot study that was used to assist in indicator selection and development of assessment methodologies.

### CS1.5.1 Purpose

The purpose of the LCWMP Pilot Study was to develop a sampling guideline that would ensure comprehensive and cost-effective evaluation of baseline conditions within the Laurel Creek Watershed. It featured exploratory analyses to determine the details of sampling programs, including minimum sample sizes, frequency of sampling, and locations of test sites.

### CS1.5.2 Data collection

The short pilot study began in autumn 1996. Sampling of watershed indicators occurred at minimum frequencies of once every 2 weeks. Each sampling unit was measured in duplicate or triplicate. The results were statistically analyzed and were used to establish appropriate sample sizes, frequencies, and locations for the System Monitoring Program (refer to Section CS1.6.1).

### CS1.5.3 Statistical analysis

A detailed statistical analysis was performed to determine the reliability of the data and enable a workup of costing scenarios based on various levels of replication and other iterations related to sampling methodology.

### CS1.5.4 Results

The environmental indicator data collected during the pilot study was submitted to the interagency team for review. Long-term recommendations for the LCWMP were established from the initial pilot study, and a recommendation was made to continue the exploratory work of the pilot study for an additional year.

## CS1.6 The program

As shown in Figure CS1.1, the LCWMP is a three-part program. The three parts, system monitoring, development monitoring, and postdevelopment monitoring, are summarized below.

## *CS1.6.1 System monitoring (watershed-wide area)*

The purpose of the system-monitoring component of the program is to characterize the baseline condition of the various environmental indicators (Figure CS1.2). This baseline then acts as a basis for comparison while and after development occurs in various areas. System monitoring then is a means of performance evaluation related to policies and best management practices for development in the watershed. Obviously, the system monitoring data are also a useful surveillance tool for overall management of Laurel Creek.

From an implementation perspective, annual system monitoring consists of the following steps:

1. Task teams (which report to the City of Waterloo throughout the year) are assembled (water quality team, aquatic habitat team, water quantity team, terrestrial features team).
2. System monitoring recommendations are reviewed, and the previous year's data are evaluated.
3. Work plans are developed based on experiences from the previous year.
4. Partnerships are established with written agreements for system-monitoring task teams.
5. System monitoring is performed throughout the year by the task teams.
6. Interim monitoring reports are prepared for each component by the relevant task teams in the middle of the monitoring year.
7. City staff review interim reports and adjust system monitoring activities where necessary (thus maximizing capacity for adaptive management).
8. Task teams continue performing spring, summer, and fall monitoring activities.
9. Task teams prepare final monitoring reports, complete with all data and recommendations for the next year.
10. City of Waterloo staff review final monitoring reports and enter all data into the LCWMP database.
11. Areas of concern within the watershed are identified, and mitigation is recommended to the appropriate City departments where necessary.

The system monitoring program undergoes complete review by the City of Waterloo every 5 years. This review looks at effectiveness of the program (appropriateness and reliability of data) and thoroughly evaluates how the information generated is used to guide other aspects of City management, particularly related to review of proposed development within the Laurel Creek Watershed.

## *CS1.6.2 Development monitoring (development-specific area)*

The development-monitoring part of the LCWMP is the responsibility of the developer and consists of three stages: Stage I, preconstruction; Stage II, construction; and Stage III, postconstruction.

The Stage I preconstruction-monitoring activities quantify attributes and general health around the development site. These baseline data assist in the design of best management practices and provide a baseline for performance evaluation over the period of development. Construction activity generally requires intrusive vegetation removal and extensive earth-grading activities, so various site-specific controls are put in place to ensure that surface water, groundwater, and other environmental features are not affected by these actions.

Stage II construction monitoring evaluates short-term flux in environmental attributes during development and is focused on assessing compliance with development controls, particularly related to sediment containment and adherence to setbacks from environmental features.

The purpose of Stage III postconstruction monitoring is to ensure that the watershed targets are being met and that the ecological health of on-site and adjacent terrestrial areas is maintained. It also indicates where, when, and whether it is necessary to perform mitigative measures and concurrently evaluates all controls in place during land development activities. This stage of monitoring extends through the 2-year period after the completion of construction.

A development industry review team was formed in August 1997 to develop a long-term development-monitoring process for the LCWMP. The team included members from the development industry, GRCA, and the City of Waterloo and still meets on a regular basis to discuss LCWMP issues.

The following sections detail each stage of the development monitoring program.

#### *CS1.6.2.1 Stage I monitoring (preconstruction)*

Throughout the preconsultation stage of the planning process, a review of existing documentation and consultation with municipal and environmental agency representatives occurs. Existing environmental features are identified for the development property and the lands immediately adjacent. Each development site is unique and may have few or a complex array of environmental features to be considered. Stage I monitoring is initiated 2 years in advance of site grading (within the City of Waterloo) so that there is a means of accounting for seasonal and annual variation while the baseline condition is defined. The City of Waterloo contacts all property owners of the lands slated for development to give development proponents due notification that pertinent approvals must be obtained and monitoring activities must be implemented before any site alterations can be made. After the preparation and submission of a monitoring program by the developer, a preconsultation meeting with City of Waterloo and GRCA staff is convened to review monitoring requirements and approve the proposed protocols. To give a regulatory backbone to the process, the Topsoil Removal Permit (which is legally required before site grading) is not issued unless the Stage I monitoring has been completed and endorsed by the City of Waterloo and the GRCA.

The results of the Stage I monitoring program are reported to the City on an annual basis by developer groups. Annual reports include a summary of the data gathered and a discussion of the implications of that information. The actual data submitted depend on the environmental features identified for monitoring but generally include the results of any surface water or groundwater chemistry analyses, surface water flows, groundwater levels, assessment of vegetation health, description of the communities of aquatic and terrestrial biota, and so on. To expedite the process and ensure consistent reporting in the watershed, the format for the submission of various data is provided by a sample Microsoft Excel spreadsheet available from City staff in digital format.

#### *CS1.6.2.2 Stage II monitoring (during construction)*

In addition to the monitoring features and protocol established in Stage I, additional site-specific features require continual monitoring through the construction period. Additional features generally consist of erosion and sediment controls, storm water management facilities, and stream rehabilitation features.

The monitoring protocol (i.e., program design) for Stage II is submitted to the City and GRCA for review before issuance of the Topsoil Removal Permit. Stage II monitoring looks specifically at the erosion and sediment controls to be implemented on site during construction and at the performance of any constructed control features (e.g., storm water management ponds or channel realignments). For that reason, Stage II monitoring starts with the beginning of construction activities on the site (i.e., topsoil removal and site grading) and extends until 90% of the tributary area to a particular facility is stabilized. On the basis of current agency (GRCA) criteria, Stage II monitoring of sediment controls takes place at least weekly. The reports for this part of the program can be tailored to the specific measures proposed for the property but are generally brief (one page), checklist-style reports that identify any measures that need repairs or replacement. Submission of results is based on two schedules. Regular monitoring reports for the assessment of erosion and sediment control measures are submitted to both the City of Waterloo and the GRCA on a weekly basis during active construction periods. During periods of no construction (often during severe winter periods for example), reports are submitted monthly. A brief annual summary of these reports is submitted to the City outlining any recurring problems with the erosion and sediment control measures and proposed means of rectifying the problem.

#### *CS1.6.2.3 Stage III monitoring (postconstruction)*

With the completion of Stage II monitoring, site construction is mostly completed, and the monitoring frequency is generally reduced to reflect the stability of the site (i.e., no more active grading, vegetation is established, site is stable). Stage III monitoring is initiated once the development reaches

90% build-out and the site is generally stable. Monitoring in this stage is a continuation of existing monitoring, but at a reduced frequency. The focus is to assess permanent mitigation measures completed on site during the construction process. Because of the extensive monitoring history of the site to this point in the development process, the stream conditions and operation and performance of any best management practices and storm water management facilities will be well known.

The Stage III monitoring generally includes the following:

- Continued surveillance of Stage I indicators
- Evaluation of stormwater management facilities
- Evaluation of stream rehabilitation works (if any)
- Assessment of the responses of terrestrial vegetative communities and aquatic communities to development

#### *CS1.6.2.4 Postdevelopment monitoring (watershed-wide area)*

Postdevelopment monitoring is the responsibility of the City and of other agencies with jurisdiction. Once the developer has completed the required monitoring for development (90% built, vegetated, with asphalt, etc.), the developer will submit recommendations for postdevelopment monitoring, the intent of which is to determine the long-term environmental effects of development. Postdevelopment monitoring activities are generally less frequent than those during the development monitoring phase because of the increased stability of the land use.

## *CS1.7 Estimated annual program budget*

The program requires up to CDN $100,000 every year from the City of Waterloo as per the breakdown given in Table CS1.4. However, with yearly recommendations and expanded future partnerships, costs to the city may

***Table CS1.4*** Estimated Annual Budget for the Laurel Creek Watershed Monitoring Program

| Laurel Creek Watershed monitoring components | Annual cost of each component (CDN $) |
|---|---|
| Water quality monitoring | 35,000 |
| Aquatic habitat monitoring | 15,000 |
| Hydrology monitoring | 18,000 |
| Terrestrial monitoring | 15,000 |
| Statistical analysis | 6,000 |
| Community education | 6,000 |
| Database development and analysis | 5,000 |
| TOTAL | 100,000 Maximum |

*Note:* Costs are exclusive of City of Waterloo staffing and do not include in-kind contributions from stakeholders and members of task teams.

be reduced. Continuation of this funding level was included in the City of Waterloo's 1998 10-year capital budget forecast; however, in 1999 it was recommended that the monitoring budget item be moved from the capital budget to the operating budget (an important step because line items in the capital budget must be reviewed annually for inclusion and are not reflective of long-term funding commitment!).

## *CS1.8 Recent program development*

The following program development action items have been under way by the City of Waterloo Development Services, Park Services, and Storm Water Management Departments in consultation with the interagency team and the development industry review team:

- A typical monitoring protocol (template) has been established for the development monitoring portion of the program in cooperation with all relevant parties.
- A process for dealing with the situations of indicator target exceedance has been developed.
- Guidelines have been developed for preparing mitigation plans, including naming the responsible parties, how the mitigation measure is to be performed, when it will take place, and by what system the effectiveness of the mitigation will be established.
- Investigative study procedures have been developed for "severely degraded" sites.
- A quality assurance and quality control manual have been written.
- A community-based monitoring program may arise from a "Pilot Project" that was performed by an external agency in partnership with the LCWMP.
- A LCWMP database has been developed and is maintained.
- Education and community awareness programs are being developed.

Yearly recommendations are made by the City of Waterloo, key stakeholders, and related team members to ensure that the LCWMP evolution is at the forefront of community and environmental monitoring and that the program continually sets groundbreaking precedents. Senior managers with the City of Waterloo are involved in distributing program information throughout the organization, which has resulted in a capacity for adaptation within each business unit.

## *CS1.9 Summary and conclusions*

The LCWS was not intended to be an endpoint but rather a springboard for action. The City of Waterloo has recognized this fact by leading the development and implementation of proactive watershed management approaches such as the LCWMP.

The City of Waterloo and all of its community watershed partners recognize that we all profit from a healthy environment. Consequently, strong efforts will continue to be made to develop public stewardship and to establish partnerships that will ensure not only that monitoring is being implemented but that there is sufficient funding available for ongoing monitoring activities and public education.

The development of the LCWMP is complete; however, adaptive management allows the program to evolve. The LCWMP is an essential part of maintaining ecosystem health for the City of Waterloo. The development and continued implementation of this program will have a positive impact on the community and the environment.

The pilot study provided valuable insight on the time requirement, associated costs, and feasibility of undertaking the monitoring program.

Monitoring the natural processes and habitats of the watershed is intended to ensure protection of the City of Waterloo's valuable resources. It also reflects a willingness to learn by doing. Monitoring provides an opportunity to note any ecological changes taking place and allows adjustments to management strategies. Monitoring is the foundation of adaptive management, which is a method of taking the perpetual feedback from the monitoring program and adjusting, refining, or rethinking goals and objectives (methods and techniques). In the Laurel Creek program, there was even capacity for adaptation within the monitoring program, as illustrated in Figure CS1.3.

The indicator monitoring plans and working plan for each part of the program are adjusted each year as per annual recommendations. Input and analysis of the data are performed regularly to ensure that the most up-to-date information is available to decision makers. Evaluation and interpretation of the data are performed a minimum of once per year and ideally should be performed after each monitoring season. Recommendations to improve the overall program and provide options for management and future development are prepared annually. A summary report is prepared annually to report the monitoring findings to the decision makers. The monitoring of the watershed is performed on a continuous long-term basis, with adjustments made as per annual recommendations.

## CS1.10 Supporting watershed documents

### CS1.10.1 The Laurel Creek Watershed Study

Monitoring is supported by the goals and objectives, targets for new development, and monitoring recommendations of the LCWS.[155] The LCWS provided direction for the formulation of the LCWMP.

### CS1.10.2 Regional policy

The Regional Official Policies Plan (ROPP) recognizes watershed planning as a "unique opportunity to understand the characteristics of ground and

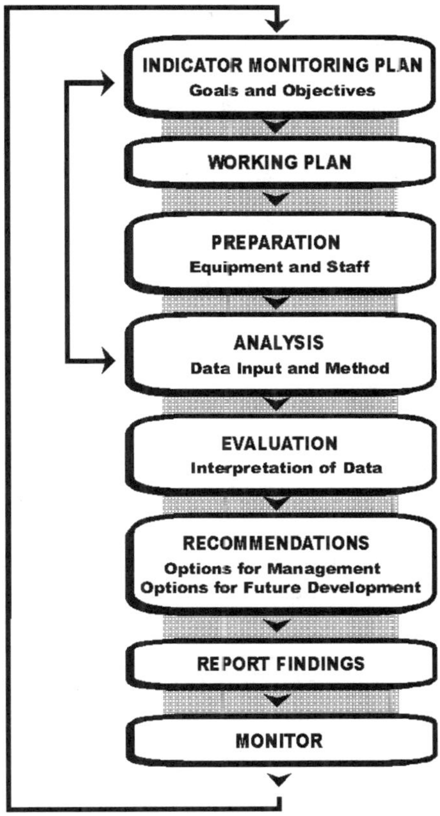


***Figure CS1.3*** Feedback loops within the watershed monitoring program implementation scheme that permit adaptation of the program based on experience.

surface water resources, evaluate interrelated natural habitats, proactively identify appropriate locations for development, and establish measures to prevent adverse impacts upon natural systems."[156]

## *CS1.10.3 Municipal policy*

The Official Plan of the City of Waterloo is a document that supports the policies of the ROPP and acts as a framework for local decision making. The plan "recognizes the Laurel Creek Watershed Study and incorporates Watershed goals, objectives and policies to guide future development and protect the environment. It is recognized that watershed and sub-watershed planning are integral to the municipal planning process."[157] One of the goals of the Official Plan is to "protect, conserve, manage and enhance natural resources including land, surface water and groundwater quantity and quality, forest and wildlife."[157] Official Plan Amendment (OPA No. 16), approved on October 14, 1993, set the stage for the LCWMP by stating the following:

> Monitoring and assessment of performance criteria, measures and mitigation procedures will be established by the City of Waterloo in consultation with the Grand River Conservation Authority, the Regional Municipality of Waterloo, the Ministry of Natural Resources and any other public agency having jurisdiction. Monitoring and assessment shall address things as hydrology, water quality, habitat conditions and constraint level area conditions. These shall be established in sub-watershed studies and implemented through corresponding district plans and other applicable documents.[157]

### CS1.10.4 District plan

District plans are a key component of the City's neighborhood planning process and contain a statement of goals, objectives, and policies that are intended to guide and direct the nature of land development within a particular community. When looking specifically at the natural environment in the District Plan for the West Side, the domain includes watershed and subwatershed areas, buffers, parks, targets, and the LCWMP. The Columbia Hills District Implementation Plan[158] is the first community plan for the West Side. Based on the City of Waterloo's Official Plan and the Council's vision for the West Side, the Columbia Hills District Plan recognizes the objective to "monitor and mitigate natural and aquatic resources including land, surface and groundwater quality and quantity, woodland, wetland and water supplies within this District."[158]

## CS1.11 Winning the 1998 Dubai International Award for Best Practices

The United Nations Centre for Human Settlements (UNCHS, Habitat) and Dubai Municipality identified LCWMP as a Global Good Practice, through the 1998 Dubai International Award for Best Practices in Improving the Living Environment.

Over 450 submissions were received by the United Nations; 124 initiatives were identified as Best Practices, and 170 initiatives were identified as Good Practices. The identification of a Good Practice has put Waterloo's initiative in the UNCHS Best Practices and Local Leadership Program. The program is a global network of organizations dedicated to sharing lessons learned from innovative practices in support of urban and community development.

The major themes of the Best Practices and Good Practices are as follows:

- Urban poverty reduction and the creation and distribution of wealth
- Urban environment and health
- Governance and civic engagement

- Disaster preparedness, mitigation, and redevelopment
- Access to shelter, land, and finance
- Status of vulnerable groups
- Gender equality and equity and social inclusion
- Use of information in decision making

The criteria for a Best Practice are poverty eradication, gender equity, and social inclusion.[159–162]

## CS1.12 Long-term goals

The long-range goals of the LCWMP are listed below:

- Determine the spatial and temporal changes in the health of the Laurel Creek Watershed's aquatic and terrestrial environment as a result of development and other human activities.
- Establish a scientifically defensible and traceable record of information (metadata) that can be readily used in land use planning, development, and management decision making.
- Develop new approaches and techniques to plan, develop, and manage the natural environment in a sustainable manner.
- Establish compliance from all parties associated with the planning, development, and management of the Laurel Creek Watershed (i.e., government, developers, builders, real estate agencies, private utilities, home owners).
- Develop public stewardship and partnerships that ensure that awareness for the necessity of monitoring is being achieved, monitoring is being implemented, and sufficient funding resources are available for ongoing monitoring activities and public education.
- Determine required mitigation measures necessary to achieve the environmental indicator targets and objectives.
- Gain assurance from partners and other involved agencies of financial backing for many years to come.

# *Case Study 2*

# *Developing an environmental monitoring program for the Uxbridge Brook Watershed*

Environmental monitoring is an essential component of the implementation of the Uxbridge Brook Watershed Plan. Performance evaluation and surveillance monitoring are being conducted throughout the watershed to evaluate the effectiveness of remedial measures and other control options to protect and maintain the integrity of the watershed ecosystem and identify any new or emerging problems. In addition, information regarding changes to existing land use and farming practices is being collected to assist in predictive water quality modeling efforts that are important for performance evaluation. The following is a description of the origins of the Uxbridge Brook Watershed Plan,[163] some of the methods used to evaluate environmental health, and the process used to develop an environmental monitoring program as part of implementing the plan.

## *CS2.1 Uxbridge Brook Watershed Plan*

The Township of Uxbridge (refer to Figure CS2.1) is located about 75 km northeast of the City of Toronto in the Lake Simcoe Watershed). The municipality has been described as a "rural Ontario treasure" and has worked hard to promote itself as an unspoiled, caring community comprising small historic towns and hamlets surrounded by a healthy natural environment. Its proximity to the greater Toronto area, combined with the beauty and tranquillity of a rural setting, has made the Township of Uxbridge an extremely desirable place to live. As a result, the municipality is under tremendous pressure to expand its existing urban boundaries.

The largest urban area within the Township is the Town of Uxbridge itself, with a population of approximately 8500 people. In 1992, the Regional Municipality of Durham announced plans to expand the existing sewage treatment plant (STP) to permit urban expansion to a target population of approximately 11,500. Projects of this type in Ontario require that a Class Environmental Assessment be conducted to determine and address any associated environmental impacts. The environmental assessment process provides the opportunity for any government agency, organization, or interested individual to voice concerns or support for the project.

Within days of the announcement of the proposed STP expansion, government agencies, interest groups, and residents were voicing their concerns to the Regional Municipality (County) and the Township. The majority of issues raised were related to perceived harmful impacts to the natural environment, especially the detrimental effects that the expansion would have on the receiving watercourse, Uxbridge Brook. The brook is considered by the community to be one of the town's natural heritage treasures and supports a healthy, cold-water brook trout fishery. Many in the community were convinced that any change to the STP would have a harmful and irrevocable impact on the brook and health of the fishery. The community's apprehension escalated when the Regional Municipality requested a deviation from provincial policy with respect to phosphorus loading to the local watercourse.

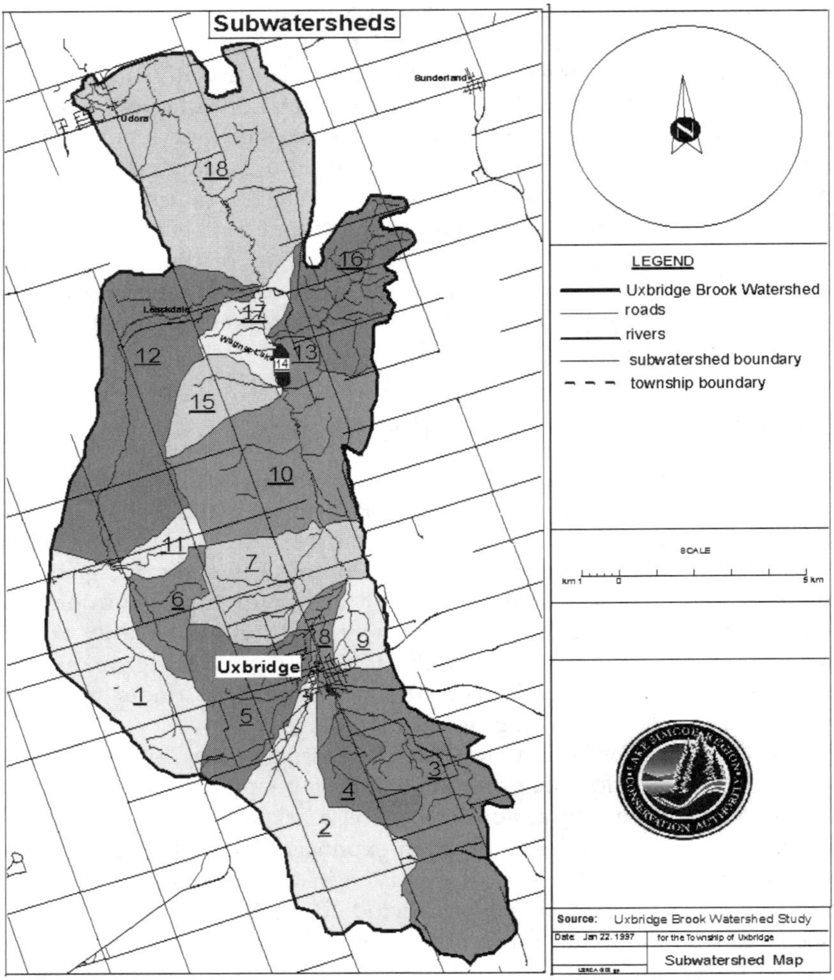


*Figure CS2.1* The Uxbridge Brook Watershed. (From Lake Simcoe Region Conservation Authority. With permission.)

Ontario's Ministry of the Environment strictly regulates water pollution control facilities in the Province. Each facility in Ontario is issued a Certificate of Approval that outlines imposed limits on the effluent concentration of certain pollutants. Phosphorus is one of the parameters regulated, and allowable effluent concentrations are determined based on consideration of the receiving water's assimilative capacity and direction from Ontario's Provincial Water Quality Objectives (PWQOs). The PWQOs set limits for a number of pollutants as a means of protecting aquatic health.[75] The PWQO for phosphorus is 0.03 mg/l, and concentrations exceeding this level are presumed to be harmful to the aquatic environment.

The Ministry of Environment has identified the Uxbridge Brook as a Policy 2 area with respect to total phosphorus. Policy 2 areas are watercourses where phosphorus concentrations are known to exceed the PWQO for more than 25% of the samples taken.[75] As a result, these rivers and streams are considered extremely sensitive, and any further inputs of phosphorus represent an unacceptable risk to health of the aquatic ecosystem. The following condition applies:[75]

> Water quality which presently does not meet the Provincial Water Quality Objectives shall not be degraded further and all practical measures shall be taken to upgrade the water quality to the Objectives.

However, as with most policy there are exceptions, and in special instances, a deviation can be requested. Policy 2 goes on to state the following:[75]

> However, it is recognised that in some circumstances, it may not be technically feasible, physically possible or socially desirable to improve water quality toward the PWQO. Accordingly, where it is clearly demonstrated that all reasonable and practical measures to attain the PWQOs have been undertaken but where:
>
> 1. PWQOs are not attainable because of natural background water quality
> 2. PWQOs are not attainable because of irreversible human-induced condition
> 3. To attain or maintain the PWQOs would result in substantial and widespread adverse economic and social impact
> 4. Suitable pollution prevention techniques are not available
>
> then deviations from this policy may be allowed, subject to the approval of the Ministry of the Environment.

The request by the Region of Durham for the deviation from Policy 2 was based on its studies, which had narrowly focused on investigating new technologies and options to upgrade the STP to reduce phosphorus concentrations in the effluent. These studies documented that the plant could not achieve the desired reduction in phosphorus necessary to achieve Policy 2 primarily because it was already operating using best-known technologies for phosphorus removal. The Region concluded that suitable pollution prevention techniques were not available and argued that a deviation was the only alternative left that would allow for further urban growth.

Although the argument made by the Region was compelling, many of the agencies and members of the community wanted answers to two other fundamental questions. Had all reasonable and practical measures to attain the PWQO for phosphorus been undertaken? Were the water quality conditions within the Uxbridge Brook irreversible? The Lake Simcoe Region Conservation Authority believed that the answers to these questions were "no." Studies conducted as part of the Lake Simcoe Environmental Management Strategy to quantify and qualify sources of phosphorus entering Lake Simcoe (see Case Study 5) provided the data to support this assertion. The Authority believed that opportunities to implement other remedial measures and phosphorus control options did exist within the watershed and that the poor water quality conditions of the Brook were reversible. The watershed agency argued that the Region had put too much emphasis on the traditional end-of-pipe solutions and not examined the big picture relative to the cumulative impacts of nonpoint source pollution on water quality within the entire watershed.

To ensure the protection of Uxbridge Brook, the Authority requested that a comprehensive watershed study be completed before any consideration could be given to a deviation in Policy 2 during expansion of the STP. The Conservation Authority maintained that a watershed planning approach would provide the big picture by examining physical, biological, and socioeconomic issues and their interrelationships. It would also furnish the necessary information to diagnose the health of the watershed and identify all the ill effects (on the other resource factors besides water quality) associated with existing and future land use practices. Once an accurate picture of watershed health was produced, alternatives to mitigate the environmental impacts associated with the proposed expansion of the STP and urban area within the Town of Uxbridge could be explored. Furthermore, a holistic strategy could be developed to recommend measures to improve water quality while ensuring the protection and rehabilitation of the other natural resources within the watershed.

The Township of Uxbridge and Region of Durham were reluctant to proceed with the expansion of the STP, knowing that such action could harm the Brook, and they agreed to the Authority's request. In 1996, the Lake Simcoe Region Conservation Authority, in partnership with the Township of Uxbridge, Regional Municipality of Durham, provincial agencies, and the watershed community began the Uxbridge Brook Watershed Plan. The Authority was appointed as the lead agency to develop the plan and did so through public consultation and consensus building. The goal of the watershed plan was simplified into two basic principles: to "protect resources that are healthy" and "rehabilitate resources that have been degraded." Consistent with the model given in this book, the overall objective of the plan was to balance future urban growth with the environmental needs of the Uxbridge Brook Watershed to ensure that no further deterioration of the sensitive ecosystem would occur. Monitoring activities played a key role throughout the plan's development by identifying impairment issues and providing the background for the comparison of potential management scenarios. This background also served as the baseline for future performance monitoring.

Throughout the entire watershed planning process, the issues relating to the expansion of urban areas and the subsequent impact on water quality and the cold-water fishery continued to dominate the community's concerns. Therefore, the remainder of this case study will concentrate on a review of the surveillance monitoring methods used to assess water quality and aquatic resources during the development of the plan. We also discuss how the results were used to develop a management strategy and delve into the ongoing performance monitoring program.

## *CS2.2 Monitoring activities used to develop the Uxbridge Brook Watershed Plan*

The first task in assessing water quality and the health of Uxbridge Brook's fishery involved reviewing existing information about the watershed to identify data gaps that would necessitate further field investigations. Heathcote[164] noted that the information to complete a detailed investigation of a watershed is seldom if ever available. Most of the past watershed planning experiences across Ontario tend to support this statement; however, significant time, money, and (in some instances) duplication of effort can be saved by contacting stakeholders and requesting their cooperation in providing information. In the case of the Uxbridge Brook Watershed Plan, a wealth of information regarding surface water quality was available; however, little fisheries, hydrogeologic, or hydrology information could be found. Land-use information and mapping were also inadequate. Given the need for this information and limited financial resources, field investigations were prioritized to fill critical data gaps.

An assessment of the condition of the surface waters of Uxbridge Brook was undertaken (primarily using historical data) to obtain an understanding of conditions for fish and aquatic organisms. Particular emphasis was placed on determining the effects of human habitation on water quality and the stream's biota to answer the question of whether the water quality of the Uxbridge Brook was a constraint on the cold-water fishery. Specifically, the ability of the watershed to retain nutrients (especially phosphorus), resist soil erosion, and trap sediment was believed to be greatly impaired because of urban land development and agricultural practices. Furthermore, warm urban stormwater runoff during the summer months was believed to be significantly altering water temperature and threatening brook trout.

Water quality information for Uxbridge Brook was available from a variety of sources and spanned a 12-year period, from 1982 to 1994. The Ontario Ministry of Environment (MOE) provided coverage for 1982–1994; the Lake Simcoe Region Conservation Authority, for 1986–1988; and private consultants, for 1991 and 1994. The most comprehensive data set was provided by the MOE and was available for two sampling sites, the locations of which are depicted in Figure CS2.2.

Historic water quality data were also available from studies completed by the Conservation Authority from 1986 to 1988[165] but were limited to nutrient and bacterial analysis. The water quality work by SENES Consultants Ltd[166] that was contracted by the Regional Municipality of Durham evaluated receiving water quality of the Uxbridge Brook for the proposed STP expansion and represented the most recent data available. See Table CS2.1 for water quality parameters evaluated during the watershed study.

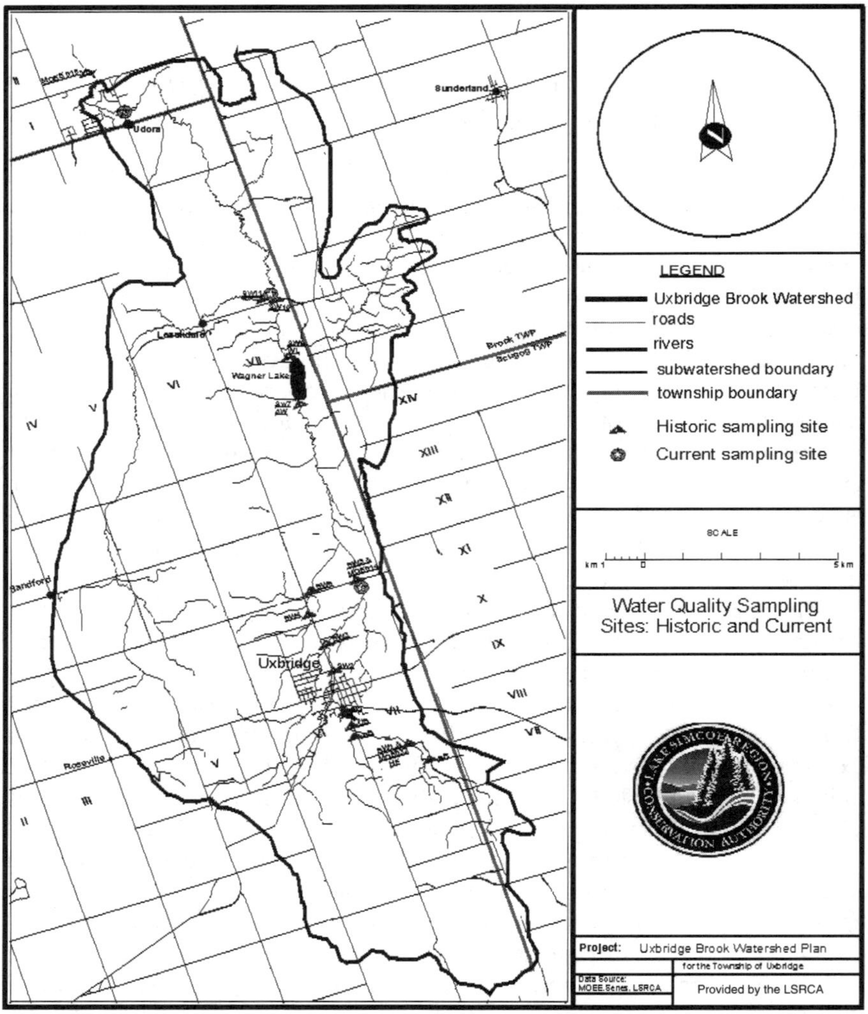


*Figure CS2.2* Current and historic water quality sampling stations in the Uxbridge Brook watershed. (From Lake Simcoe Region Conservation Authority. With permission.)

***Table CS2.1*** Water Quality Parameters of Interest in the Uxbridge Brook Watershed Study

| Chemical | Physical | Biological |
|---|---|---|
| Total phosphorus | Conductivity | Total coliform |
| Filtered reactive phosphorus | Turbidity | *Escherichia coli* |
| Nitrates | Suspended solids | Fecal streptococcus |
| Nitrites | pH | *Pseudomonas aeruginosa* |
| Total Kjeldahl nitrogen | Alkalinity | |
| Ammonia | Temperature | |
| Phenols | | |
| Chloride | | |
| Sodium | | |
| Copper | | |
| Mercury | | |

*Source:* Adapted from Lake Simcoe Conservation Authority (LSRCA), Uxbridge Brook Watershed Plan, LSRCA, Newmarket, Ontario, 1997.

The first phase of the water quality analysis involved a comparison of the raw data to the PWQO[75] to identify where sample concentrations exceeded allowable limits. In the instances in which a parameter was observed to exceed a PWQO, a more extensive analysis examining spatial and temporal trends was completed. Results from these investigations were routinely depicted at public meetings using an intuitive "high, low, close" graph, with a clearly marked line identifying the PWQO. Water quality results were also compared with the information collected as part of the aquatic inventories to ascertain whether the parameters identified were in fact having a detrimental effect on the fishery. For example, copper can potentially be toxic to fish at concentrations exceeding 0.005 mg/l. Concentrations of copper were found to slightly exceed the PWQO for 18% of all samples recorded at station MOE032 (see Figure CS2.2). To determine the effect that the copper concentrations might be having on the health of the fishery, a survey to evaluate the MOE032 reach of the Brook was conducted. Results of the fish community inventory found this site to be one of the most healthy and productive reaches within the entire watershed; therefore, it was concluded that the copper loading was not limiting to the cold-water fishery at this particular site. This comparative technique was conducted for each of the stations at which a PWQO parameter had been exceeded to accomplish the following:

- Better understand the relationship between water quality and the fishery
- Establish management issues
- Identify future monitoring needs

In the case of water quality parameters in exceedance of PWQOs, the feasibility of control options to reduce contamination was investigated. In the case of copper, no human source was identified, so no control measures were considered. However, continued surveillance monitoring of copper was recommended because the potential for harming the aquatic resource remained and the source was unknown.

As indicated above, extensive field investigations were conducted to collect the necessary information regarding the state of aquatic resources. Data collected through inventory work included information on the fish community and water temperature. Efforts were also made to identify areas providing significant aquatic habitat, such as reeds and nursery. These data were compiled using a geographic information system (GIS) to update a data layer for environmentally significant areas. GIS was further used as part of a constraint analysis exercise to limit urban development within environmentally sensitive areas. Any human land use activities observed to be degrading habitat were also recorded for future reference when developing the aquatic management component of the watershed plan.

Fish community sampling was conducted throughout the watershed and at all of the sites at which historical data existed or water quality sampling was initiated during the study. Historical and new stream stations were evaluated through a qualitative survey methodology designed to document the presence or absence of various species, particularly cold-water habitat indicators such as sculpin (*Cottus* sp.) and brook trout (*Salvelinus fontinalis*). Quantitative surveys were also completed at several stations using a single-pass methodology.[167]

Stream temperature was evaluated using two different techniques. Changes in temperature between different reaches of stream were evaluated using a roving spot-temperature survey. Sampling days and times were determined by the criteria outlined by Stoneman and Jones[168] in their rapid temperature assessment methodology. Stream temperatures were also evaluated at several additional stations using Stowaway™ dataloggers.

Like-reach habitat evaluations recorded information on stream gradient, substrate, riparian vegetation characteristics, stream shading, and adjacent land use. Human activities observed to be altering or destroying fish habitat (i.e., on-line ponds, water taking, channel alterations) were noted and entered into a database for future reference or action. All of the data collected were entered into a GIS for future reference.

Results of the water quality monitoring and fisheries inventory were somewhat predictable in that they confirmed earlier suspicions that the Uxbridge Brook ecosystem had been severely impacted by changing land use and human activities within the watershed. Undoubtedly, the most consequential and controversial conclusion drawn from the monitoring results was that water quality and the future of the cold-water aquatic resource would be at risk should the expansion of the STP and the urban area be allowed to occur, based on present-day conditions. Two water quality parameters were found to be limiting aquatic health: temperature and (to a lesser degree) phosphorus. On the basis of these conclusions, stakeholders recommended the development of resource targets for the protection of water quality and aquatic biota, as listed in Table CS2.2.[163]

***Table CS2.2*** Uxbridge Brook Watershed Plan Resource Targets

| Resource | Target |
|---|---|
| Water quality | To protect ground and surface waters that currently meet the Ministry of Environment and Energy Provincial Water Quality Objectives and Ontario Drinking Water Objectives; to enhance below-objective ground and surface waters to at least meet these objectives, recognizing that in certain areas exceptional water may require more stringent criteria and that in some situations natural conditions may make attaining this goal undesirable |
| Aquatic resource | To maintain and, where possible, improve aquatic habitat of Uxbridge Brook to ensure the continued and improved health of the aquatic ecosystem |

*Source:* Adapted from Lake Simcoe Conservation Authority (LSRCA), Uxbridge Brook Watershed Plan, LSRCA, Newmarket, Ontario, 1997.

Fortunately, opportunities to improve water quality and rehabilitate the aquatic resource had also been identified within the Uxbridge Brook Watershed. A comprehensive review of urban and rural human land use practices as part of the monitoring activities provided the information to develop a comprehensive water resource management strategy as part of the watershed plan. Included were recommendations regarding the establishment of new development policies and bylaws, an environmental constraint map, and a schedule for the completion of remedial measures and rehabilitation projects to achieve the resource targets.

One of the best examples of how data collected from the field investigations were used is the development of an overall phosphorus control strategy. Not surprisingly, this exercise received the most attention from the stakeholders, especially developers and the watershed community. The goal of the strategy was to reduce phosphorus loads entering the brook by 25% and to reduce phosphorus concentrations within the brook to the PWQO of 0.03 mg/l.[75] On the basis of the intent of the resource targets, this goal had to be achieved and maintained before municipal plans to expand the STP and continue urban growth could proceed.

To determine whether the resource target was achievable, an adaptation of a predictive water quality model known as HYDROSIM was used to evaluate various management and growth scenarios. The model, developed by BEAK Consultants Ltd,[169] is a modified version of the U.S. Environmental Protection Agency's Hydrological Simulation Program—Fortran (HSP-F) model.[143,144] Once calibrated using observed monitoring data, it can be used to predict changes in runoff water quality associated with the implementation of best management practices (BMPs) or changes in land use. The model uses GIS to develop a spatial representation of the watershed in which it simulates runoff and pollution transport. The model output provides the user with daily, monthly, or annual phosphorus loadings expressed in units of mass per unit time. Loadings for the Uxbridge Brook analysis were reported in kilograms per year and were calculated from urban and rural areas for point and nonpoint sources. To determine whether the resource target was achievable for phosphorus, four model scenarios were evaluated as follows:

- Scenario A—baseline conditions; do nothing
- Scenario B—baseline conditions; implementing rural and urban BMPs
- Scenario C—future growth (official plan designation); do nothing
- Scenario D—future growth (official plan designation); all BMPs

Scenario A represented the present-day phosphorus loadings from the Uxbridge Brook and was used as the benchmark when evaluating the different management scenarios. Phosphorus loadings for all four of the modeling scenarios are listed in Table CS2.3 and depicted graphically in Figure CS2.3.

Total phosphorus loadings for Scenario A were estimated at 5446 kg/year. Rural sources such as runoff from livestock operations and cropland soil erosion were the two largest contributors, accounting for more than 53% (2887 kg/year) of the total load. Natural sources were the next most significant contributor, providing roughly 25% (1358 kg/year) of the total load. Urban sources accounted for the remaining 23% (1241 kg/year) of the total phosphorus load, with stormwater runoff the largest urban contributor.

Scenario B involved the implementation of a full range of BMPs to address urban storm water, septic systems, runoff from livestock operations, and cropland erosion. The total phosphorus load from this scenario is estimated at 3817 kg/year and, with a reduction of 1669 kg/year from implementation of BMPs, is well within the 25% loading target. Modeling showed that the largest reductions in loading could be achieved by implementing rural BMPs associated with livestock sources and cropland erosion. Loading reductions from urban sources accounted for a disappointing 130 kg/year

***Table CS2.3*** Summary of Phosphorus Loadings Based on Modeling Scenarios

| Loading component | Phosphorus loads (kg/year) by modeling scenario | | | |
|---|---|---|---|---|
| | A | B | C | D |
| Septic systems | 191 | 95 | 228 | 132 |
| Sewage treatment plant | 110 | 110 | 285 | 285 |
| Urban storm water runoff | 940 | 810 | 1041 | 876 |
| Agriculture—livestock | 1528 | 764 | 1528 | 764 |
| Agriculture—cropland erosion | 1359 | 680 | 1359 | 680 |
| Pasture or fallow | 559 | 559 | 559 | 559 |
| Forest | 492 | 492 | 492 | 492 |
| Wetland | 150 | 150 | 150 | 150 |
| Scrubland | 157 | 157 | 157 | 157 |
| TOTALS | 5486 | 3817 | 5799 | 4095 |

*Source:* Adapted from Lake Simcoe Conservation Authority (LSRCA), Uxbridge Brook Watershed Plan, LSRCA, Newmarket, Ontario, 1997.

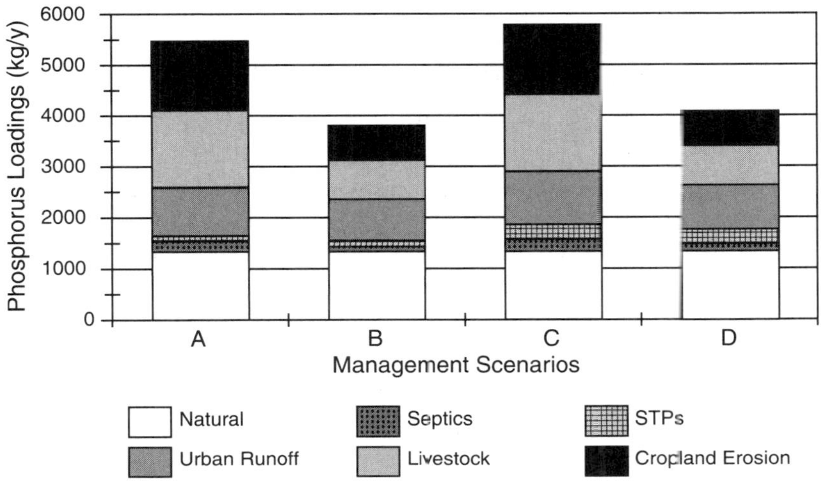


***Figure CS2.3*** Modeling results for Uxbridge Brook management scenarios. (From the Lake Simcoe Region Conservation Authority (LSRCA), Uxbridge Brook Watershed Plan, LSRCA, Newmarket, Ontario, 1997. With permission.)

because of the lack of opportunity to construct adequate storm water management facilities in some of the existing urban areas.

Scenario C represents the worst-case scenario for the Uxbridge Brook, based on uncontrolled urban growth to accommodate an additional 4000 people and no implementation of BMPs. The total loading to the brook was estimated at 5799 kg/year, an increase of 313 kg/year above the existing conditions. This estimate assumed significant increases in phosphorus loadings associated with the following:

- Additional septic systems to accommodate a 1000-person increase in rural population (approximately 37 kg/year)
- Expansion of the STP to accommodate 4000 additional urban residents (approximately 175 kg/year)
- Urban stormwater runoff from new, impervious development areas (101 kg/year)

Additional inputs from agricultural activities were not expected and therefore not predicted. Scenario C was justifiably unacceptable to the stakeholders.

Scenario D depicted the projected phosphorus load from future growth as described in Scenario C but included the implementation of all BMPs discussed in Scenario B. The total predicted load for Scenario D was 4095 kg/year, representing a total reduction of 1391 kg/year from Scenario A, the base condition. This reduced phosphorus loading would achieve the water quality target but require the full implementation of all urban and rural BMPs. Any reductions associated with urban BMPs would be more than compensated for by the additional loadings related to future growth.

Results from the modeling analysis indicated that the proposed expansion to the STP and urban area could proceed without further impairment of the Brook's water quality as long as the following were implemented:

- A full range of water quality BMPs to achieve the desired water quality target
- Stringent water quality protection guidelines for all new development within the Town of Uxbridge
- BMPs to improve water quality and water quality protection guidelines for urban development before further expansion to the STP or urban area would be allowed
- An environmental monitoring program to document improvements associated with the implementation of remedial measures, to provide a capacity for adaptive management, and to identify any new or emerging environmental problems within the watershed

Throughout the development of the Uxbridge Brook Watershed Plan, it was essential that monitoring information was provided to all stakeholders to ensure that decisions regarding the future health of the ecosystem could be assured. The phosphorus management strategy was one of many proposed management activities developed as part of the watershed plan. It is an example of how monitoring information can be applied to develop policy and initiate action. In this instance, resource targets could be achieved while accommodating the socioeconomic needs of the community—of course, action was needed and it could only come at a price.

## *CS2.3 Monitoring implementation of the Uxbridge Brook Watershed Plan*

In 1998, the Township of Uxbridge Council formally adopted the Uxbridge Brook Watershed Plan by passing a resolution recommending that the results of the report be incorporated into future municipal planning documents (most notably a proposed secondary plan) and that efforts to implement the rehabilitation strategy be initiated immediately. It was also recommended that an environmental monitoring program be established to collect physical, chemical, and biological data to measure the effectiveness of adopted remedial measures and facilitate response to any future environmental problems.

The creation of a Watershed Committee of Council that was fully endorsed by the Township of Uxbridge was the next step in the implementation process. Comprising community volunteers, staff from the various agencies, and council members from both the Region and the Township, the committee was established to direct rehabilitation activities. The Lake Simcoe Region Conservation Authority was designated by the Committee as the lead agency responsible for managing the proposed activities and reported directly to the Watershed Committee. The Watershed Committee in turn reported back to the Township of Uxbridge Council, presenting annual updates on the status of the implementation program, work plans, expenditures and budget requests, and proposals for program development. The Committees' main objectives were to do the following:

- Develop annual work plans and budgets to direct implementation efforts
- Build awareness within the community about the significance of the Uxbridge Brook ecosystem relative to the well-being of the community
- Mobilize the community to action through education and opportunities for involvement (thereby generating a network of volunteers, with direct spin-offs to the bottom line of the project)
- Solicit and maintain involvement of provincial agencies, interest groups, and other stakeholders
- Raise funds and develop partnerships to implement capital projects
- Design and implement an environmental monitoring program

The design and development of an environmental monitoring program was not considered an initial priority by the Uxbridge Brook Watershed Committee. It was recognized that most of the water quality monitoring activities undertaken as part of the watershed plan were being continued by the Conservation Authority to enable the detection of long-term trends. Efforts to establish resource protection policies and develop a rehabilitation program to address resource management issues were expected first. For this reason, the development of a comprehensive environmental monitoring program was only recently initiated, in 2000.

The monitoring program was developed using the same public-participatory approach used in the development of the watershed plan, with decisions made through consensus. A working group was established including representation from the following agencies: the Federal Department of Fisheries and Oceans; the Ontario Ministries of Natural Resources, Environment, and Agriculture, Food and Rural Affairs; the Regional Municipality of Durham, the Township of Uxbridge; and the Lake Simcoe Region Conservation Authority. Additional stakeholders from local organizations and the development community were also invited to participate. A number of criteria were established at the outset of discussions to assist members in the development of program objectives, a selection of which is listed below:

- Monitoring initiatives should be related to the resource targets established within the watershed plan.
- Scientific assessment and analytical methods must be based on strong science and be defendable (i.e., water quality sample analysis must be conducted by an accredited MOE laboratory).
- Results must be relevant and used for decision making to assist the committee in adapting rehabilitation programs and development policies to achieve and maintain the resource targets.

Given these criteria, the goal of the Uxbridge Brook Watershed Environmental Monitoring Program is to collect information to identify changes in ecosystem health and to ensure that action is taken to achieve and maintain the resource targets for the Uxbridge Brook Watershed. Both surveillance and performance monitoring methods were included in the monitoring program. Surveillance monitoring (as discussed in Chapter 3) is conducted to obtain long-term trends and to identify emerging problems. Performance monitoring evaluates management activities and is triggered by a change in land use or management practices. It is primarily conducted at the site level.

## *CS2.3.1 Surveillance monitoring activities*

Three different approaches to surveillance monitoring are part of the Uxbridge Brook implementation program. The PWQOs[75] were adopted as the water quality targets, so surveillance monitoring stations for water quality were established at three sites within the watershed (Figure CS2.2). Two of the sites have a sampling history, whereas the third (located at the Village of Udora) is new and only instrumented in the summer of 2001. Analytical water quality parameters include those previously sampled by the MOE as part of the Provincial Water Quality Network so that results can continue to be compared with the PWQOs.[75] Sampling is conducted monthly, with the exception of nutrients, suspended solids, pH, dissolved oxygen, and temperature, which are also sampled on a precipitation event basis. Water quantity (flow) is also being measured at each of the sites to facilitate the calculation of annual loadings so that progress toward the loading target of 25% for phosphorus can be assessed. This information is essential for compliance purposes, given watershed planning recommendations for improved water quality before continued urban expansion.

In addition to water quality, biological surveillance monitoring of both benthic macroinvertebrates and the fish community was initiated at the sites sampled during the development of the watershed plan. Benthic macroinvertebrate sampling was not included during the initial assessment of watershed health but has been considered an essential indicator to predict the future health of the Uxbridge Brook aquatic system. A multivariate, reference-condition approach (see Section 3.3.2.2) will be tested in 2001 and, depending upon the cost and difficulty, may be adopted as the standard sampling approach. As discussed in Chapter 3, the ongoing debate about which method to use in Ontario (i.e., univariate or BioMAP vs. multivariate or reference condition) has resulted in a delay in selecting a specific technique. For surveillance purposes, a sampling frequency of once every 3 years was recommended.

Fish community sampling was conducted during the watershed planning process and data collection will continue; however, a new method, the Ministry of Natural Resources' Stream Assessment Protocol,[85] will be used to determine the presence or absence of indicator fish species and their relative distribution. Stream temperature monitoring is also being conducted; however, sampling sites have been reduced for financial reasons. Only sampling locations in the headwaters of the brook and specific reaches displaying marginal conditions in terms of cold-water habitat will be monitored. The fish community sampling will follow the same 3-year pattern prescribed for the benthic macroinvertebrate sampling. Temperature monitoring will be conducted during the June 1 to October 1 period for the next 3 years, using the same sites and methods practiced in the development of the watershed plan.

The last type of surveillance monitoring being conducted involves the collection of information regarding land use and management practices. This information is essential to provide many of the input variables for the HYDROSIM model used during the watershed planning process to estimate phosphorus loadings and their relative source contributions within the watershed. Changes in land use and management practices have to be updated on a regular basis for the model to accurately predict phosphorus loads. To ensure that this occurs, the Township of Uxbridge has accepted responsibility for notifying the Authority regarding any changes in urban or rural land use within the watershed. Once notified, the Authority is responsible for updating the GIS data layers.

The Authority is also obliged to survey the farm community regarding their farming practices once every 5 years to gauge uptake of educational programs and assess changes to past practices. A survey was conducted in 1996 as part of the original watershed planning process; therefore, another survey was scheduled for the winter of 2001. Last, the Authority is at present involved in the delivery of a Clean Water Incentive Program in partnership with the Township of Uxbridge and the Watershed Committee. The program provides technical and financial assistance to landowners willing to undertake environmental projects to improve water quality within the watershed. Information regarding the work being completed is also entered into the database and into the model to predict the water quality benefit associated with the remedial work.

### *CS2.3.2 Performance monitoring*

Performance monitoring is at present being conducted at the site level within the watershed and usually involves the evaluation of a particular BMP or change in land use to assess the impact on water quality and biota. Conventional techniques such as upstream and downstream sampling, paired subwatershed approaches, and runoff collection methods have been employed in both rural and urban areas to assess performance of individual BMPs and the impacts of poor management practices.

Strict water quality guidelines established for urban areas as part of a detailed study completed by TSH and Associates[170] have necessitated that performance monitoring be conducted for all new stormwater management facilities in the watershed.

The purpose of the study was to address stormwater runoff from future development and to further refine the stormwater management strategy contained in the watershed plan to retrofit existing urban areas within the Town of Uxbridge. Guidelines approved by council in 2000 require that all new developments achieve a 90% reduction in phosphorus removal. Increased water temperature was also recognized as a concern, and a new guideline has proposed that there shall be "no net increase in water temperature of marginal or cold water reaches between June 1 and October 1 as a result of stormwater discharge from new development."[170] If approved, this will necessitate the combination of a number of urban storm water BMPs including infiltration (or exfiltration) measures to achieve this level of control.

Monitoring during the pre- and postconstruction period (up to the time that the facility is assumed by the municipality) is the responsibility of the developer and is a condition stated in servicing agreements between developers and the Township of Uxbridge. The municipality will assume the monitoring of the facility from the property developer, subject to all the conditions being met in the servicing agreement.

A monitoring protocol to assess performance of new urban stormwater facilities is at present being developed but was not available at the time of writing. Existing monitoring programs developed by the City of Waterloo (see Case Study 1) and by the City of Mississauga are being reviewed.

### *CS2.3.3 Cost recovery, data management, and communication*

Partners involved in contributing financially to the monitoring program are limited to the Township of Uxbridge, the Lake Simcoe Region Conservation Authority, and the Ontario Ministry of the Environment. Plans to expand the financial partnership, along with opportunities to involve local schools and nearby colleges and universities, will be explored in the future.

Data management of all the information collected as part of the Uxbridge Brook Environmental Monitoring Program is the responsibility of the Lake Simcoe Region Conservation Authority. Information is being stored in database form and is incorporated into the Authority's GIS to facilitate analysis and reporting. To promote awareness, information collected through the program is provided free of charge to any interested parties.

Data reports are produced and presented to the Township of Uxbridge council each and every year to coincide with the development of work plans and budgets for the upcoming year. All stakeholders involved in the monitoring program receive a copy of the report along with an updated data compact disc. Plans to place these annual reports along with the data files on the Township of Uxbridge and Lake Simcoe Region Conservation Authority Web sites are also in progress so that the data can be downloaded at any time. To review the progress achieved over the last 5 years of implementation of the Uxbridge Brook Watershed Plan, a public symposium has been scheduled for 2002. This forum will educate the community regarding implementation activities; provide an opportunity to obtain community feedback; and secure support for the continuation of existing implementation and monitoring programs or the development of new initiatives, if necessary. In this way, the monitoring program will help ensure that the principles of the Uxbridge Brook Watershed Plan—to protect resources that are healthy and rehabilitate resources that have been degraded—will be achieved.

# Case Study 3

# Integrated Stormwater and Watershed Management System® — An emerging tool for watershed health planning and monitoring

## CS3.1 Introduction

Freshwater ecosystems have fueled human development and civilization for more than 5000 years. They give us clean drinking water, fish to eat, and opportunities for recreation. We harness them to generate renewable energy and remove human waste, irrigate agricultural fields, and carry ships to port. Everywhere on Earth, from the smallest village to the largest metropolis, the lives of people are intimately intertwined with fresh, and often flowing, water. Our transformation of the natural world to serve our water demands is impressive. In the United States, for example, only 2% of rivers and streams remain free flowing and undeveloped. As alluded to at the June 2000 annual meeting of the Canadian Institute of Planners in Prince Edward Island, existing dams can hold approximately 60% of the entire annual flow of our nations' rivers in any year. By the end of the decade, it is predicted that two thirds of the world's stream flow will be regulated for human purposes.

As discussed in Chapters 1 through 5 of this book, water availability (and hence watershed health) is vulnerable to climate change and variability, in addition to the myriad of ways in which humans affect it. The major connecting link in a watershed ecosystem is the flow of water. The total quantity and distribution of flow is the water balance (see Figure CS3.1).

Water balance is linked to the hydrologic cycle (see Figure CS3.2), which describes the process of water inflow from precipitation and the outflow of water by evapotranspiration, groundwater recharge, and stream flow. This process forms a dynamic balance that can vary with time depending greatly upon climatic conditions—as is evident in many countries around the world.

The water budget, or how much water is available in each stage of the cycle, is critical in that it dictates the conditions of an ecosystem. As our watershed health definition indicates, human activities are part of this ecosystem, and conflicts arise when water use exceeds the resources available in terms of the water budget conditions. A balanced water budget and healthy ecology are intertwined. This includes the partitioning of water into terrestrial zones and in sufficient quantities so that human uses and ecological functions can occur. How, where, and how much water flows determines the quality of the water at any time and resulting potential for flooding, the shape and stability of stream banks, the health and diversity of vegetation, and the availability of fish and wildlife habitat.

As human use of a watershed or region increases, anthropogenic needs (which are often ecological stressors) can change the water budget. Changes to the water budget cause direct and significant changes to the above resources. These changes can be extremely alarming (in the case of hazard floods and droughts) and will reduce the ability of the human population to use and enjoy the resources of the watershed.

Canada has been viewed as an area with an abundance of fresh water. This is not necessarily the case in some regions because water is not

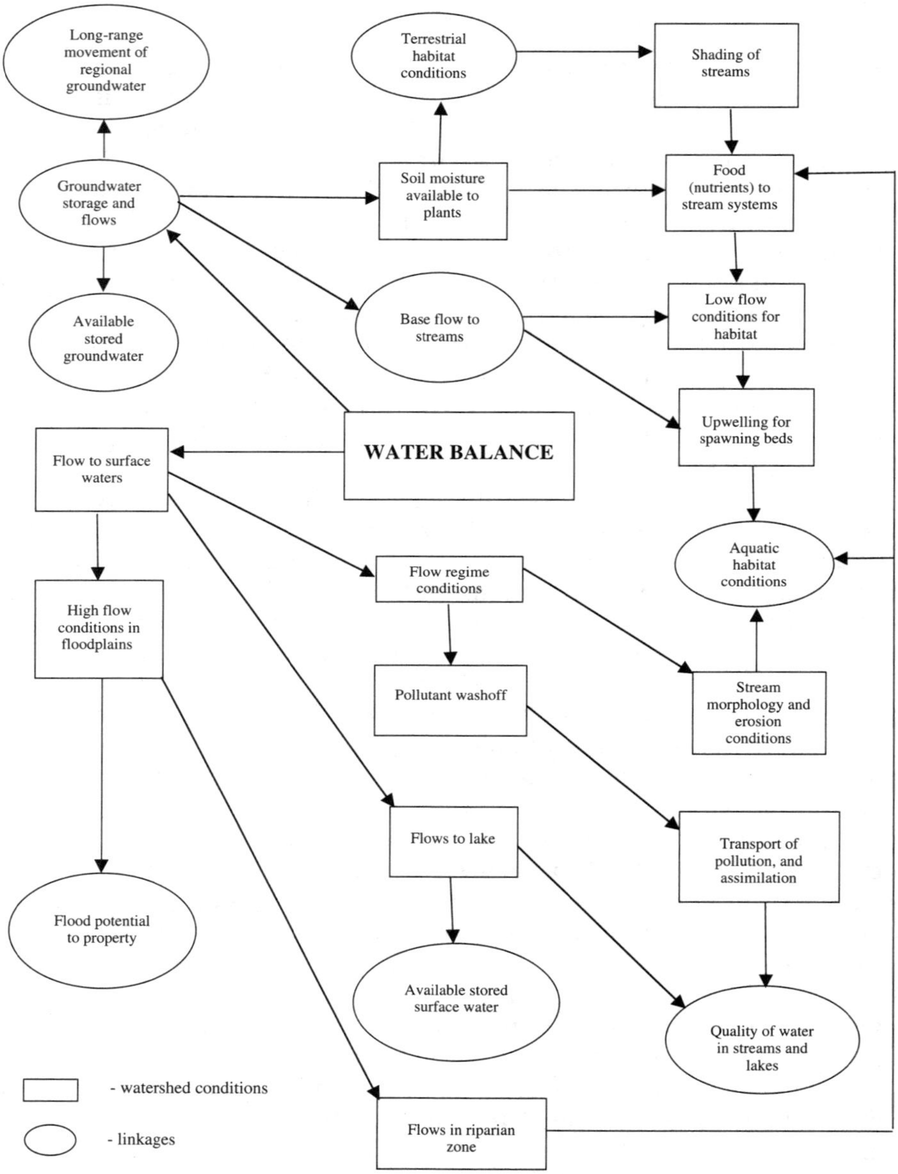


***Figure CS3.1*** ISWMS water balance model pathways. (Courtesy of Greenland International Consulting Inc.)

evenly distributed and is not available where it is needed. For example, significant flooding concerns, perhaps as a result of global warming factors, have now arisen because of extreme flash or extreme storage-related

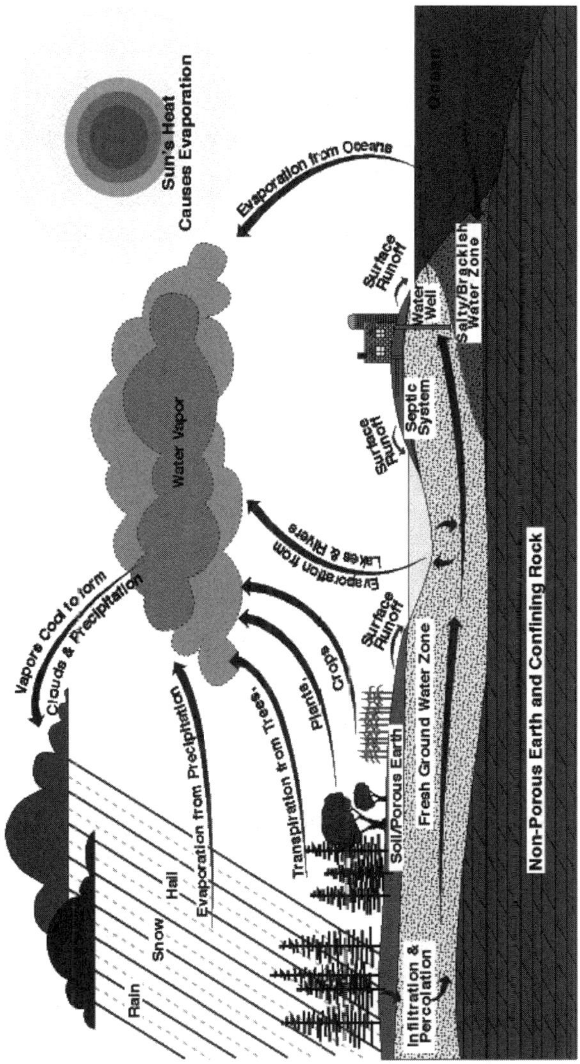


*Figure CS3.2* Hydrologic cycle. (Courtesy of Greenland International Consulting Inc.)

flood events, at one end of the spectrum, and insufficient supply to match consumptive uses, at the other end. In terms of resolving watershed health problems, increasing attention is now being paid to the water budget and to implications for allocating waters from various zones toward both human and ecological uses (e.g., potable water supplies, stream base flow, irrigation waters, recreational and/or hydroelectric reservoirs, etc.). Over the past decade, for example, serious concerns have arisen in communities that rely on groundwater for municipal use because of growing populations and the impact of water extraction on ecosystem sustainability. In May 2000, concerns across North America about watershed health, particularly related to groundwater quality, were heightened after the *Escherichia coli* bacteria outbreak in Walkerton, Ontario—the worst groundwater tragedy in North America. After agricultural runoff contaminated the local aquifer, 7 people died and 2300 became sick from the Walkerton tragedy.

Nutrient management (surface water) and groundwater quality concerns, impacts on local ecosystem sustainability by groundwater extraction for global export, and other water use pressures from development (both urban and rural) now exist in the Blue Mountain and Beaver Valley region, east of Walkerton. This includes water storage and the impact of stream flow changes on the ecosystem for the region's hydropower facilities. In addition, agricultural water use demands are increasing dramatically as a result of shifting to dwarf apple trees with shallower root systems that require irrigation. In fact, about 25% of Ontario's apples are grown in the Blue Mountain and Beaver Valley region. The sensitivity of this region's ecosystem compounds these concerns because environmental characteristics are controlled by current water balance conditions. In addition, the limestone bedrock and complex topography of this region mean that local streams were historically base flow rich and supported a number of (increasingly rare) species that demand cool, relatively constant temperatures, including important fisheries resources, such as naturalized trout and salmon populations.

The region is also part of the Niagara Escarpment, which is designated by the United Nations Educational, Scientific, and Cultural Organization as a World Biosphere Reserve. Intuitively and as our watershed health definition suggests, depletion of groundwater tables and changes in stream flow patterns will change habitat conditions and alter species' (including humans) distribution and perhaps reduce the number of species that can live in the Blue Mountain and Beaver Valley region. Two examples are presented below that highlight principles of our Closed-Loop Watershed Health Monitoring Model (as described in Chapters 2 through 5). This case study also describes the potential of an integrated information and hydrologic modeling software system to assist with preparation of watershed and subwatershed management plans and to contribute to related surveillance and performance monitoring initiatives.

## CS3.2 Developing subwatershed plans and monitoring programs using the Integrated Stormwater and Watershed Management System

In this section, we describe how the Integrated Stormwater and Watershed Management System (ISWMS) software can be used in subwatershed planning and monitoring programs.

### CS3.2.1 Objectives and case study locations

Watershed planning is an effective approach for managing land use change. In the strictest sense, it is not a tool to assess land use change, although some of this may come into play as part of performance evaluation. Subwatershed planning and the development of management strategies have long been recognized as important tools in developing an effective means of managing environmental conditions and our natural resources.[171]

We are learning that because of the complexity of the hydrologic cycle in the Blue Mountain and Beaver Valley region (see Figure CS3.3), current water use activities have far-reaching impacts. Typically, watershed boundaries are defined according to the surficial drainage basin. However, because groundwater flow systems typically occur within aquifers (local, shallow or regional, deep systems), which do not follow the surface water boundaries, the groundwatershed can be considerably larger than the surface water basin. In other words, groundwater extractions from aquifers in one watershed may in fact affect adjacent watersheds. Therefore, a watershed health management strategy, such as a subwatershed plan, using our scientific assessment techniques from Chapter 3, provides for the ability to investigate environmental processes on a broad scale within a reasonable (watershed or regional) boundary. There are still some cross-boundary issues to deal with in developing the strategy, however, such as groundwater system linkages, water use allocation, and cumulative withdrawal impacts, and so on, between adjacent municipalities or governments. Watersheds, however, provide the geographical basis for analyzing and quantifying the environmental and social processes that govern ecosystem health.

The Blue Mountain and Beaver Valley region is the headwaters of several major Ontario river systems, including the Grand, the Nottawasaga, the Saugeen, and the Beaver. Groundwater discharge supports stream base flow and, in headwater areas such as Black Ash Creek, high-quality cold-water stream habitat. The local aquifer also supports a diverse range of agricultural, commercial, and recreational uses and provides drinking water supplies for the majority of the region's 15,000 residents. In 1998, the ability of the groundwater system to meet demands placed on it was called into question when parts of the region were declared a disaster area (under federal and provincial legislation) as a drought dried up wells and affected livestock and crop operations. Against this backdrop and with

*Figure CS3.3* "Four season" water needs within the Blue Mountains and Beaver River Valley, Ontario, Canada. (Courtesy of Greenland International Consulting Inc.)

low water conditions again in 1999, as well as proposed increases in groundwater use (specifically by water bottling operations), concerns were raised by residents that the local aquifer would not be able to sustain current groundwater uses plus new demands. Of particular concern was the export of water from the area, especially when drought conditions affected rural wells and historical base flow minima were being measured in streams and rivers for the second year in a row.

To resolve this challenging dilemma, Greenland International Consulting Inc. from Toronto, Ontario was retained to gain an understanding of the role and function of the physical, hydrologic, hydrogeologic, and ecological systems of all watersheds within the Blue Mountain and Beaver Valley region. At the same time, Greenland International, in partnership with the local Conservation Authority, also prepared a detailed subwatershed plan for the Black Ash Creek system, which is located within the same region. Integrated computer modeling and analysis technology were used, including MODFLOW™ and ArcINFO™ software, as well as Greenland International's new ISWMS computer program. These tools were used to develop targets and management plans, with a focus on the creation of Living Documents, which incorporated performance evaluation monitoring components. In addition, the overall watershed health monitoring program, which was developed from ISWMS computer model simulations using hydrologic parameters and scoped Tier 2 assessment techniques (refer to Chapter 3), included schedules for periodic review and amendment over the long term. The Black Ash Creek Subwatershed Plan was completed in 2000, and the Blue Mountain and Beaver Valley Groundwater Management Plan was completed in 2001.

The ISWMS-based subwatershed and groundwater management plans also included an interactive Web site and emphasis on public education, using an integrated two-track process of communications and involvement that was consistent with our recommendations from Chapter 4. The Web site had an information management system (IMS) function for the project team, including (1) a structured database to maintain nonspatial information such as monitoring results, textual information, and georeferencing data; (2) digital map files to maintain spatial data such as land use, hydrologic units, and surficial geology; and (3) a geographic information system (GIS) to provide the linkage for team members between the spatial and nonspatial data. The IMS was used throughout the project to provide a dynamic mapping base; to retrieve, analyze, and display data; and to evaluate the interactions between the multiplicity of factors affecting water management. Owing to seasonal (winter) constraints and the somewhat remote community location, the IMS was also used as a tool during a very successful public consultation process, along with more traditional public information and stakeholder workshops.

A partnering model, developed previously by Sustainable Development & Monitoring Inc. of Waterloo, Ontario, was also modified and implemented for the groundwater management investigations to bridge the gap between the study and long-term implementation by forming community (and political) links and partnerships. To achieve this study objective, our political linkage methods from Chapter 2 were used. Finally, both management plans integrated water balance and groundwater computer modeling data, in conjunction with scientific assessment techniques from Chapter 3, with remote sensing technologies to monitor watershed health during and after the projects. Further details are presented later in this case study.

## *CS3.2.2 Overview of ISWMS software*

In 1998, Greenland International began developing an integrated, GIS-based, real-time hydrologic model, in partnership with the Nottawasaga Valley Conservation Authority (NVCA) from Ontario, to address flood forecasting

*Figure CS3.4* ISWMS software by Greenland International Consulting Inc. (Courtesy of Greenland International Consulting Inc.)

needs of this watershed agency. During the project, the NVCA and Greenland International recognized a need to deal with large amounts of watershed data (such as stream flow, biological attributes, flood-prone areas, etc.) and saw an opportunity to integrate data for multiple watershed management and monitoring objectives. Using the software, for example, streamflow data is used for hydrologic, water balance, and hydraulic model calibrations; stream flows and water quality at multiple locations can be predicted based on flow and water chemistry information from a limited number of sites; and watershed and stream health indicators can be tied to stream flow and water quality information for reporting for performance evaluation.

Greenland International subsequently developed an integrated and modular software program with a flexible approach to data management and multiple analyses of watershed and drainage system conditions. The first version of ISWMS (see Figure CS3.4) software was completed in 2000 and has been used to develop subwatershed plans in conjunction with scientific assessment protocols and other aspects of the Closed-Loop Model. In particular, this software has direct uses in the watershed management and watershed health monitoring fields, including the following functions for both rural and urban settings:

- Flood forecasting
- Water quality and nutrient management
- Master drainage planning and stormwater management
- Water balance maintenance
- Uncertainty analysis

The integrated approach used to develop ISWMS (version 1) not only enhances the modeling functionality for watershed planning that is aimed at sustainable development, ecological dynamic stability, and multimanagement objectives but also significantly improves the efficiency of data management and execution of model simulation. Application of the software for subwatershed planning and groundwater management projects in the Blue Mountain and Beaver Valley region (which is discussed later on) also strengthened political linkages and community education and involvement to complete the closed-loop approach in the study area. To illustrate the political involvement aspect of the study, note that Greenland International was invited near the end of the groundwater management project to present its unique ecosystem study approach to the legislature of the Ontario government. This in turn was very important for the project's management committee to be able to make a long-term commitment to watershed health monitoring (for at least the next decade) and to participate further in future development phases of ISWMS by Greenland International.

ISWMS (version 1) is now able to integrate several database management system software and GIS technologies with accepted hydrologic and hydraulic simulation techniques from the U.S. Environmental Protection Agency (EPA) SWMM and EXTRAN models, as well as the most popular planning-level models from Canada. The model also has water quality, water balance, and uncertainty modeling capabilities. ISWMS has the ability to simulate the following hydrologic components:

- Interception and surface storage
- Precipitation
- Snow cover and snowmelt
- Evapotranspiration
- Infiltration
- Groundwater and base flow
- Overland runoff, channel routing, minor and major drainage systems, reservoirs, and stormwater management facilities

Similar to another hydrologic model, Visual OTTHYMO™ (a hydrologic and storm water management program), that was developed by Greenland International, ISWMS is not a conventional command-based program. Instead, it is an object-based software package, whereby icons represent watershed elements such as catchments, watercourses, reservoirs, and so on, when developing a watershed simulation model. The following overview is presented so that ISWMS users can become mindful of the software's modeling regimes to determine its applications and limitations within the context of developing watershed management plans and watershed health monitoring programs.

The development of ISWMS divided the software programming components into two logical units: front-end and back-end programs. The front-end programming involves the overall design of the program including database management, interactions with the back end, and general software architecture. The front end is often referred to as the "user interface programming." The front-end programming was completed using Microsoft Visual Basic®, a programming language recommended in the literature for the development of Microsoft Windows® applications that make use of SWMM computational routines.

The back-end programming involved the development of the hydrologic and hydraulic computational routines. Most of the back-end routines had already been written in FORTRAN and were available as public domain software. This included the U.S. EPA SWMM and EXTRAN models. Selected routines from OTTHYMO and QUALHYMO planning models used in Canada were also incorporated.

Internally, data storage and data manipulation in ISWMS are achieved through a system of Microsoft Access® databases. Inputs include hydrologic and hydraulic parameters necessary to define the drainage system, plus other simulation parameters necessary to conduct detailed design or planning-level evaluations. In addition, the database system stores the results of the analysis, field monitoring measurements, and other data specified by the user that may be relevant to watershed management. For example, the user is able to store water quality and biological data collected at specific nodes in the drainage or conveyance system for future use in developing a watershed health monitoring program.

In some cases, users may choose to use remote data, stored in other databases, to maintain consistency and avoid duplication. For example, any part of the hydrologic data can be stored within a remote GIS. Consistency in the data types and structure must be maintained in both systems to ensure compatibility with ISWMS modeling and analysis requirements.

One of the most frequent applications of ISWMS is in flood forecasting and warning. Each year, flooding causes large societal losses in terms of property damage and human health. Flood forecasting is automated in ISWMS through a seven-step wizard, as follows:

1. Initialization of system state variables
2. Recalibration
3. Climate forecasting
4. Flood forecasting
5. Flood-vulnerable area (FVA) identification
6. Flood contingency or emergency and action plans
7. Forensic analysis

The system can also be used to predict property damage and assess risk to life from severe storm events by inputting multiple-climate station data, simulating stream runoff, and calibrating results with available monitoring data.

The major objective of FVA management in ISWMS (or through any other method, for that matter) is to identify structures in the floodplain that are affected by water level elevations associated with a specific storm event. Three databases and integrated links to flood-forecasting components are included in ISWMS. The first database provides rating curves (water level vs. flow) at each identified cross section of the watercourse in the watershed. The rating curve database can usually be established by running a backwater model (e.g., HEC-RAS) in the watershed. The second database records geo-referenced locations of vulnerable man-made structures and land use obtained from flood risk maps and field surveys. Through database manipulation and querying in ISWMS, information can be determined about flooded structures, such as the type of structure, its location, and its associated land use. The third database relates to flood warning as it stores the appropriate contact and recommended action plans when a flood warning is to be issued. The three databases are internally linked and can be easily maintained and upgraded in Microsoft Access (or other common database programs).

ISWMS is also a useful tool for water balance analysis and stormwater management. Water budget determination implies a quantification of the volumes of storage and rates of water movement from one physical state and location to another. Water balance models can be very useful when establishing the long-term water budget components (see Figure CS3.1) of an area before urban development and the anticipated changes in the hydrologic cycle (see Figure CS3.2) after development. They are also a vital precursor to establishing water allocation decision frameworks. The accounting is made of water entering, leaving, and remaining in storage during a specified time period over a study area, which can be defined by topographic, political, or other arbitrary criteria. The water components considered in the ISWMS modeling approach include precipitation in the form of rainfall or snowfall, rainfall runoff, snowmelt, upper soil infiltration, evapotranspiration, and groundwater recharge. Results of induced recharge (through implementation of best management practices such as infiltration trenches, perforated sewer pipes, reduced lot grading, etc.) can also be analyzed as a measure of the performance of the development process from a watershed health point of view. The ISWMS water balance modeling output can be used to establish relative proportions of predevelopment and postdevelopment water budget components. ISWMS provides great flexibility when assessing different development configurations and control measures to achieve specific recharge targets.

During uncertainty analysis, ISWMS can be applied under situations with and without watershed health monitoring data. When there are observed climate and stream flow data, for example, the model needs to be calibrated and verified before its application. The classical procedure for model calibration and verification often runs into two major problems, namely, that (1) the modeled output successfully matched the observed output for some period of the observation but fails to do so for other periods of record and (2) model parameter sets that vary over a large range may result in indistinguishable model outputs. The classical procedure also has severe limitations to properly dealing with the problems. At best, the consequences could be dismissal of modeling results because of skepticism. At worst, wrong decisions could result from ignorance of model limitations and assumptions. The problems of the classical procedure originate from the negligence of inherent modeling uncertainty and parameter identification. As such, a number of authors[172–174] have been advocating the emergence of a new and powerful paradigm of model application that must explicitly consider the issues of modeling uncertainty and parameter identification.

If application of the model is opted for in the absence of observed data, practitioners face further difficulties in determining parameter values and the uncertainty associated with model predictions. Therefore, it is practically desirable to have a model equipped with a toolkit that provides a guideline for the selection of parameters, enables sensitivity analysis, and estimates the model predictive uncertainty. The uncertainty-analysis component in ISWMS is thus aimed at enhancing the success of model application by helping practitioners explicitly deal with the inherent modeling uncertainty and parameter identification in situations with and without observation data. This function of ISWMS is also valuable for developing watershed health monitoring programs, in conjunction with scientific assessment techniques from Chapter 3, when climate change and global warming at the watershed level are concerns.

### *CS3.2.3 Applying ISWMS in the Blue Mountain and Beaver Valley region, Ontario*

The Blue Mountain and Beaver Valley region is an area of the Niagara Escarpment (a World Biosphere Reserve) in Ontario that drains to Georgian Bay and Lake Huron. Surface water and groundwater management are key issues in the development of watershed plans and stream health monitoring strategies. In this area, concerns regarding water use and availability relate to the potential depletion or degradation of water as a result of competition over the resource. This includes cumulative impacts from the issuance of Permits to Take Water by the Ontario government. Generally, a permit is very easy to obtain in Ontario, with conditions (such as monitoring) rarely applying. Public hearings are also not necessary for these permits, nor are direct notices to adjacent landowners. Permit issuance and details (i.e., rate) are not based on sustainable yield through knowledge of local water budgets. Unfortunately, however, this approach gained widespread acceptance in the late 1900s, when information was not widely available. A compliance monitoring approach or subwatershed-based review of permit applications was not used. In Ontario, water rights (other than for personal use, agricultural use, fire fighting, etc.) are based on common law, such that landowners have the right to "reasonable" use of water on their properties as long as extraction does not have an adverse effect on adjacent landowners. Beyond permit woes, there is a general sentiment in the area that population growth could outpace supply.

Unequal distribution of water resources in the Blue Mountain and Beaver Valley region has led to political pressure in favor of diversions for water supply (which has led to further outcry related to potential for erosion, lack of attention to sustainable development, and problems with waste assimilation in receiving-water bodies). Accompanying concerns over water availability and the potential for aquifer mining (a situation in which extraction exceeds recharge) are significant for stream health because groundwater is inseparably linked to surface water, and reduced quality or quantity may have impacts on local fisheries and other recreational and cultural resources.

#### *CS3.2.3.1 Watershed management and flood forecasting*

Black Ash Creek is a cold-water fishery that drains to Collingwood Harbour, in Collingwood, Ontario. Remedial action plan (RAP) studies by the Government of Canada in the 1980s recognized the importance for ensuring sustainable development within the basin, given that the system outlets to Collingwood Harbour, the only currently delisted RAP site by the Government of Canada.

The watershed (see Figure CS3.5) drains 3260 hectares and originates above the Niagara Escarpment. It receives groundwater in its headwaters because of the porous soils, spring seepage from fissures in the face of the escarpment, and steep gradients. In general, land use within the watershed is predominantly rural and agricultural. However, the headwaters include a large ski resort and holdings slated for golf course development. The downstream basin area within the only urban center, Collingwood, is the most developed but also contains large rural areas, including another golf

***Figure CS3.5*** Black Ash Creek Watershed, Ontario, Canada — Satellite image: September 1999. (Courtesy of Greenland International Consulting Inc.)

course, as well as industrial and commercial shopping malls and residential developments.

The Black Ash Creek Subwatershed Management Plan was prepared by the NVCA and Greenland International in July 2000 to detail the drainage basin's natural resources, develop a stormwater management plan, identify stream rehabilitation opportunities, and ensure environmental protection as urban development proceeds within the municipality. This was the first time that ISWMS was implemented for a subwatershed planning project in the Blue Mountain and Beaver Valley region. The subwatershed plan provided a comprehensive inventory of natural heritage features; identified a green-space system, which will be protected and enhanced as development proceeds; and proposed a long-term watershed health monitoring program that was consistent with our Chapter 3 recommendations. Using ISWMS, the plan also provided an impact assessment of future development on the natural heritage features as well as a management plan to guide protection of the green-space system and future development within subwatersheds of Black Ash Creek.

*CS3.2.3.2 Groundwater management and water balance monitoring* ISWMS was also used in the Blue Mountain and Beaver Valley region to develop a groundwater management plan (see Figure CS3.6). To do so, continuous hydrologic simulations were run during summer and winter

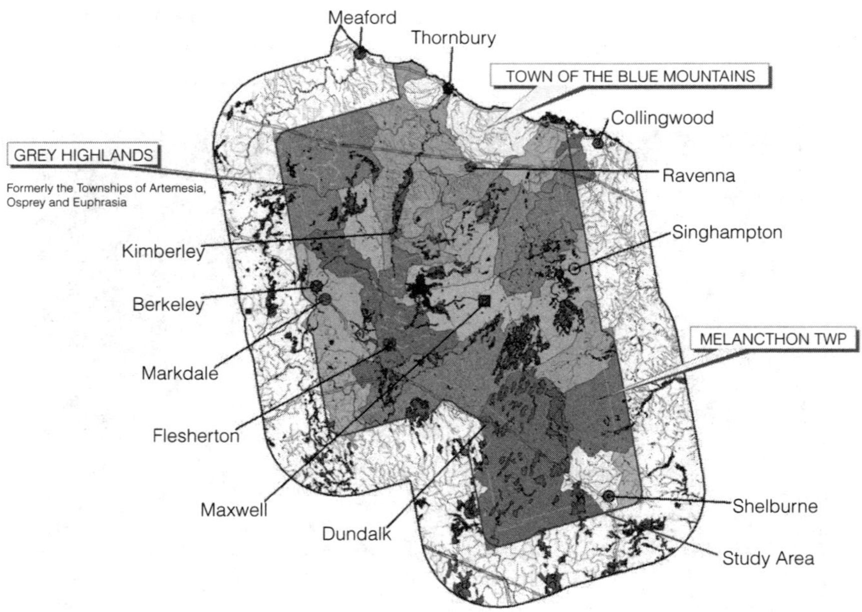


***Figure CS3.6*** Watersheds of the Blue Mountain and Beaver River Valley, Ontario, Canada. (Courtesy of Greenland International Consulting Inc.)

conditions, and surficial infiltration capabilities were estimated (both of which are essential water balance considerations). Calibration for the ISWMS approach relied on climate station data as well as on groundwater level and stream flow monitoring data that were measured during the study, from January 2000 to April 2001.

Figure CS3.7 presents an ISWMS computer screen image of various modeled watersheds within the Blue Mountain and Beaver Valley region. ISWMS data integration used an ArcINFO database management system. The hydrologic output data from the ISWMS models (i.e., infiltration parameters) were integrated directly into MODFLOW for hydrogeologic analysis. These watershed models were then used to generate various graphical representations of the groundwater system, including aquifer, groundwater flow, and bedrock topography maps; cross-sections; and overburden thickness maps.

Finally, the integrated ISWMS and MODFLOW models by Greenland International were Internet-enabled to permit public input and facilitate the study's long-term partnering model. As emerging technologies, both the Internet-enabled models and the partnering model represent important ways to reduce long-term operational model costs; enhance and maintain political, public, and industry partnerships; update all watershed models with data

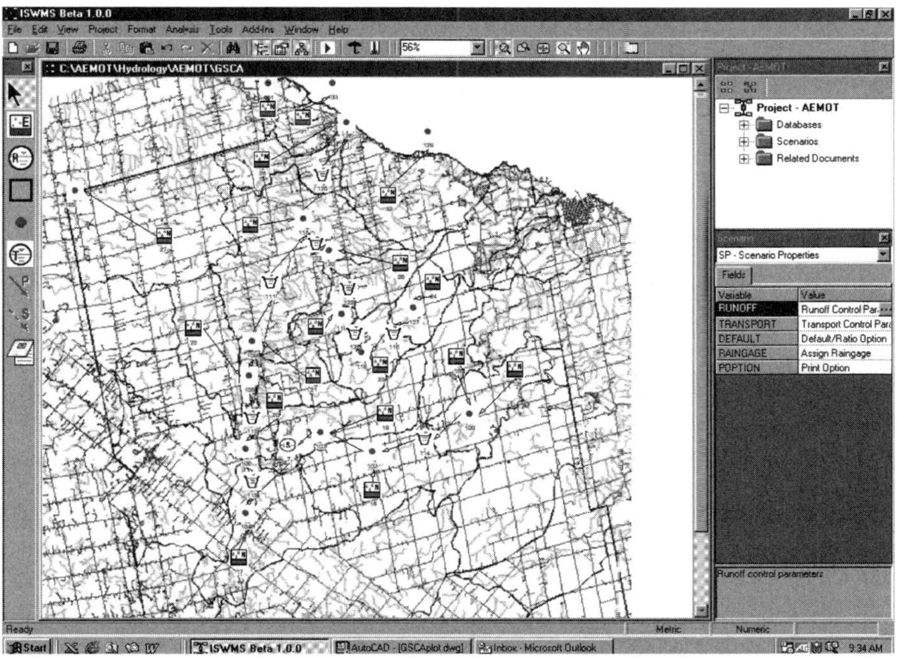


***Figure CS3.7*** Computer screen view of an ISWMS simulation for watersheds within the Blue Mountains and Beaver River Valley, Ontario, Canada. (Courtesy of Greenland International Consulting Inc.)

from the field-monitoring network; and hence ensure an effective watershed health-monitoring program. This integrated IMS- and Web-based approach included automated code for acquiring, uploading, and formatting files and subsequently storing information in a database. The surface and groundwater models by Greenland International can now be run for flood forecasting and other water management objectives, in conjunction with a monitoring network that will gather climate, stream flow, and groundwater-level data across the region.

The ISWMS models for the Blue Mountain and Beaver Valley region will also allow local governments to undertake more detailed investigations (i.e., similar to those of the Black Ash Creek study) as part of future subwatershed plans in support of new development or water use demands.

## CS3.3 Conclusions

Greenland International Consulting Inc. identified the need to have a comprehensive management tool to systematically deal with multiple watershed management objectives, particularly those issues of immediate concern related to human health and safety and the health of watershed ecosystems. To this end, ISWMS was developed by integrating several emerging information software technologies with accepted hydrologic and hydraulic simulation techniques. The software is now being used in Canada for projects by Greenland International and can also be applied anywhere in the world. In addition, ISWMS operates on a database management engine using either internal or external data warehouses, such as existing GIS systems or other databases. A remote data access technology will make use of existing data stored in agencies. Tailoring will meet specific functional needs as well as data format availability. Proposed development phases for ISWMS include the addition of an environmental-monitoring module (e.g., using measured stream bioassessment and water chemistry data, as well as scientific assessment techniques from Chapter 3) and a stormwater management pond maintenance module, using Greenland International's sediment removal and construction site research data from Canada. Greenland International also embarked in 2001 on a quest to integrate weather radar data, groundwater modeling software, distributed runoff modeling, and remote-sensing data with the first version of ISWMS.

Several conclusions have been identified from the ISWMS development program as an emerging tool for watershed planning and management, synthesis, hypothesis generation, and reporting on the status of watershed health, namely, the following:

- Integration of database management and hydrotechnical computer models makes stormwater management and watershed planning enhanced, simpler, environmentally sound, and cost effective.
- Integration of accepted computational hydrologic models gives ISWMS worldwide applicability in terms of resolving water use conflicts and predicting impacts caused by climate change.

- The capability of ISWMS to produce identical results with accepted hydrologic models makes it trustworthy and user-friendly for government agencies, consultants, universities, and research institutions.
- Incorporation of an application procedure considering modeling uncertainty makes ISWMS unique.

In terms of applications, the ISWMS model can be used to examine remedial measures for enhancing or restoring watershed health through integration with monitoring programs. The model is now being used for flood hazard management. Model application for subwatershed plans will contribute to the business climate for ISWMS in terms of ensuring sustainable community-based water use strategies.

Model application for subwatershed plans will reduce or defer municipal capital water and sewage infrastructure costs (e.g., through maintenance of stream or river base flow for wastewater assimilation), and ISWMS development phases are tailored to addressing government policies and priorities in environmental sustainability through the creation of alliances or partnerships between both rural and urban stakeholders.

## CS3.4 Case study project participants

The following parties contributed to the application of ISWMS within the study area: Greenland International Consulting Inc., Concord; Nottawasaga Valley Conservation Authority, Angus; Sustainable Development and Monitoring Inc., Waterloo; 4DM Inc., Toronto; and Kije Sipi Ltd, Ottawa, Ontario, Canada.

# Case Study 4

# Advanced remote sensing technologies for watershed health monitoring

## *CS4.1 Introduction*

In the absence of human activities, the characteristics of habitats used by all biota are determined primarily by physical features and natural processes. The magnitude and rates of change of habitat variables are therefore determined largely by geology and climate. When human activities are introduced into natural areas that were previously devoid of human activities, impacts can cover the range from direct and obvious to indirect and subtle. Despite our best efforts to establish, for example, a watershed management plan, the strategy will be based on a finite set of information and assumptions about development timing, build-out period, and human activity. Consequently, it is important that surveillance and performance evaluation be undertaken to enable adaptive management.

In remote areas, careful attention must be made to develop an appropriate set of monitoring indicators and corresponding adaptive management measures. These measures would then be employed if monitoring results indicate that adjustments are required to the management measures already in place. A good monitoring strategy for accessible and remote areas is based on an understanding of the goals to be achieved. Further, triggers for adapting the plan in response to monitoring data should be established. Very often, the funding for monitoring is restricted, and therefore proposals for monitoring that require special training and intensive inventory are seldom useful. Therefore, it is especially important in remote areas to develop a monitoring strategy that will be inexpensive to implement and that will provide some insight into the success of the plan and into ways in which to adapt to unexpected results. Two examples are presented here that highlight principles from earlier chapters while highlighting the potential of advanced technological systems for surveillance monitoring, including remote sensing and the Internet.

Recent advancements in remote-sensing technologies are providing unique opportunities to better manage watershed resources. These systems are also being used to monitor potential environmental hazards that impact social and economic well-being. A monitoring program can also encourage schools, organizations, and citizens to participate in the collection of data. Recent advances in protocols and procedures with geospatial data processing have included assembling collected data in a centralized Web site database that is accessible by all watershed stakeholders. Figure CS4.1 presents a format that was used for a regional groundwater management study in Canada,[175] as discussed in Case Study 3.

The focal point of the information management system (IMS) shown in Figure CS4.1 is an Internet data interface. The Internet-enabled system included (1) a structured database to maintain nonspatial information such as climate, groundwater, and stream flow monitoring results, textual information, and georeferencing data; (2) digital map files to maintain spatial data such as land use, hydrologic units, and surficial geology; and

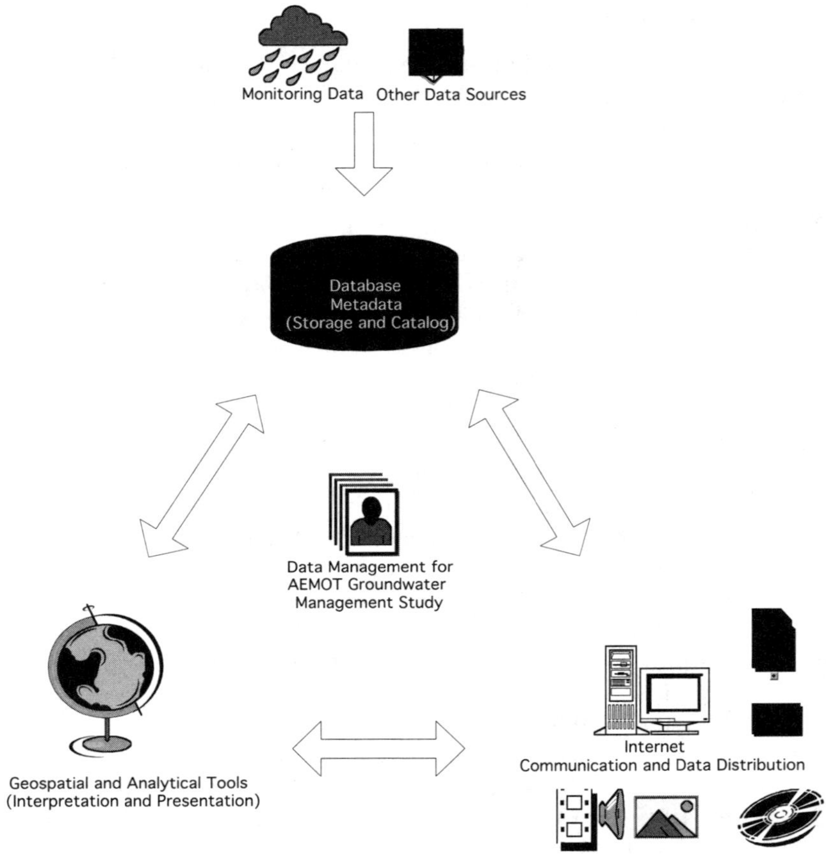


***Figure CS4.1*** Advanced information management system for the AEMOT groundwater study in Ontario, Canada. (Courtesy of 4DM Inc. and Greenland International Consulting Inc.)

a geographic information system (GIS) to provide the linkage between the spatial and nonspatial data. Finally, the IMS shown in Figure CS4.1 was also used as a tool for a public consultation process that was consistent with our Chapter 4 recommendations.

*Remote sensing* refers to any data collection methodology that does not require direct human observation. Aerial data measurements represented the first step toward reducing the amount of labor needed to gather geospatial information. Such methodologies have also helped pinpoint areas (and reduce associated costs) for biophysical ground measurements and on-site testing of geophysical phenomena. Remote sensing has yielded highly precise elevation data and extremely high resolution site data. Digital cameras and global positioning system (GPS) receivers are making aerial data easier to process and ever more accurate.

## CS4.2 Recent advances in remote sensing

Satellite data became commercially available in 1972 when the U.S. government launched the first LANDSAT satellite, which was specifically designed to provide imagery of the Earth. LANDSAT 1 had a spatial resolution of 90 m, meaning that objects of that size and larger could be distinguished. However, this low resolution restricted the use of the data to government agencies such as the U.S. Environmental Protection Agency and was therefore not accepted by private-sector engineers and planners as being a practical tool for site-level monitoring.

Recent advances mean that software can now handle large image data files that can be georeferenced using GPS and a plethora of new sensing technologies. For example, in January 2000, satellite imagery resolution of 1 m became available to those outside of the intelligence community from the IKONOS satellite, which was developed by the Space Imaging Corporation of Thornton, Colorado in the United States. Figures CS4.2 and CS4.3 illustrate images of two landmark structures in the United States captured using IKONOS technology.

In spring 1999, the U.S. government lofted LANDSAT 7, which offered imagery of the highest resolution and lowest price of any LANDSAT. Recently, *airborne light detecting and ranging* (or LIDAR) sensors offer some of the most accurate elevation data in the shortest time span ever by bouncing laser beams off the ground.

## CS4.3 Use of radar satellite data for watershed health monitoring

### CS4.3.1 Overview

RADARSAT-1 is Canada's first Earth-observation satellite and currently the world's most advanced commercial radar satellite. A research project[176] was undertaken to assess how soil moisture mapping could be obtained using RADARSAT-1 data to reduce the cost of ground measurements and increase the accuracy of quantifying the watershed's hydrologic budget. Also, this approach permitted a rationalization of the current soil moisture–monitoring system with associated reductions in data collection costs and increased water resources benefits. Furthermore, the pilot project examined the potential long-term benefits of incorporating RADARSAT-1 data in operational flood-forecasting requirements.

The project's technical objective was to assess the feasibility of mapping soil moisture using RADARSAT-1 data at the field level, but more important, at the basin level. RADARSAT-1-derived soil moisture mapping was also assessed by Jobin[176] for use with a distributed hydrologic model in order to assess operational benefits at the user level. Essentially, the optimization of monitoring hydrologic variables was based on using RADARSAT-1 imagery that was calibrated using field-level moisture probes and grab sampling.

*Figure CS4.2* Washington Monument: September 30, 1999. The first commercial 1-m resolution image from IKONOS. (From Space Imaging Corporation. With permission.)

Furthermore, because the field observations were relatively sparse and could not represent all land and soil conditions, the RADARSAT-1 images were correlated spatially against distributed antecedent precipitation index (API) maps using weather radar precipitation data. As a final step, the developed moisture model was tested at the watershed level by integration with a distributed hydrologic model.

A significant component of this project was to assess the value and feasibility of soil moisture mapping using RADARSAT-1 imagery data. To complete this assessment, an extensive field-level soil moisture data–monitoring program was established in parallel with the RADARSAT-1 imagery acquisitions. The elements of this program included (1) deploying up to four soil moisture probes at selected environmental monitoring stations, (2) locating appropriate soil moisture sampling sites throughout the basin, and (3) conducting a soil moisture–sampling program in parallel with the RADARSAT-1 imagery acquisitions.

*Figure CS4.3* Hoover Dam: December 25, 1999. 1-m resolution image from IKONOS. (From Space Imaging Corporation. With permission.)

## CS4.3.2 Location of pilot project watershed

The Mississippi River Watershed was chosen as the pilot basin for the soil moisture study. It encompasses more than 4000 km$^2$ of land area and is located immediately west of Ottawa, Ontario (see Figure CS4.4). The river is a major tributary to the Ottawa River and is roughly rectangular in shape, stretching 100 km long by 40 km wide. The main stem is more than 200 km in length and drops more than 320 m in elevation, with an average slope of 0.6 m/km. There are more than 250 lakes, with numerous wetlands that are mainly concentrated in the upper half or in the western parts of the basin. The watershed presents a variety of land uses, including a large fraction of forested lands with small tracts of agricultural and urban areas that are mainly concentrated in the lower eastern sections.

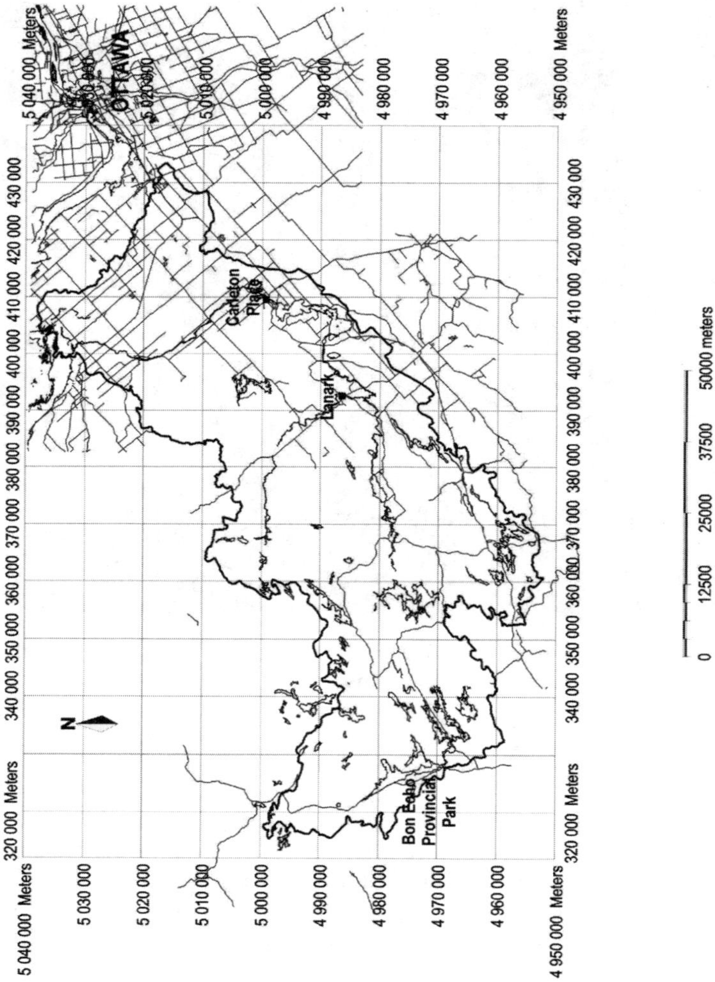


*Figure CS4.4* Location of the Mississippi River Valley Watershed, Ontario, Canada. (From Kije Sipi Ltd. With permission.)

## *CS4.3.3 Methodology*

A total of 15 RADARSAT-1 images were acquired during the course of the study. The intent was to cover a range of seasonal conditions. All imagery was acquired in ascending orbits (south to north). This meant that the satellite imagery data were acquired over the basin at approximately 18:00 hours on any given day. The selection of this particular orbit avoided potential periods of dew on the ground that could occur on the descending orbits (06:00). Past research had indicated that soil moisture mapping was likely difficult if not impossible under forested cover. This is due to the scattering effects of the forest cover on the microwave signal. Considering this constraint, Jobin[176] selected only sites with open areas as potential candidates for sampling. Candidate sampling sites were identified using a recent false-colored LANDSAT image. Subsequently, two field visits and one air reconnaissance trip were completed to identify the final locations for soil moisture sampling. A color video camera, mounted in nadir position, was used to record the ground track during the air reconnaissance mission. Both field and air photographs were also taken during these field trips. This photographic information proved to be invaluable in assessing the final sampling locations.

From the initial air reconnaissance survey, most candidate sites were retained, but a few were substituted for others nearby to make use of the existing snow course survey stations. Most remaining candidate sampling locations were visited three times and assessed in terms of accessibility, vegetation cover, and overall terrain and soil characteristics. Permission to access the land also had to be secured from the landowners. A total of 12 main sites covering the entire basin were isolated by Jobin[176] using this selection process. Also, soil moisture sampling was conducted near the locations of the soil moisture probes. Altogether, 16 sites were sampled and surveyed using global positioning systems during the course of the project. Furthermore, during the field reconnaissance, five sampling stations were established at each of the main 12 sampling sites. The sampling stations had to be at least 15 m apart and in the same type of microterrain. The location of each of the 60 sampling stations was georeferenced against local landmarks to ensure that measurements could be made at the same locations from sampling session to sampling session.

## *CS4.3.4 Conclusions*

Jobin[176] drew several observations and conclusions from this initial attempt at mapping soil moisture using the derived relations with the RADARSAT-1 data. As illustrated in Figure CS4.5, soil moisture generally increased uniformly going from east to west and ranged from 10 to 80%.

The total range of soil moistures calculated using the remotely sensed RADARSAT-1 imagery may have been overestimated and should have been less than 45% (according to ground-truthing work over the sampling period).

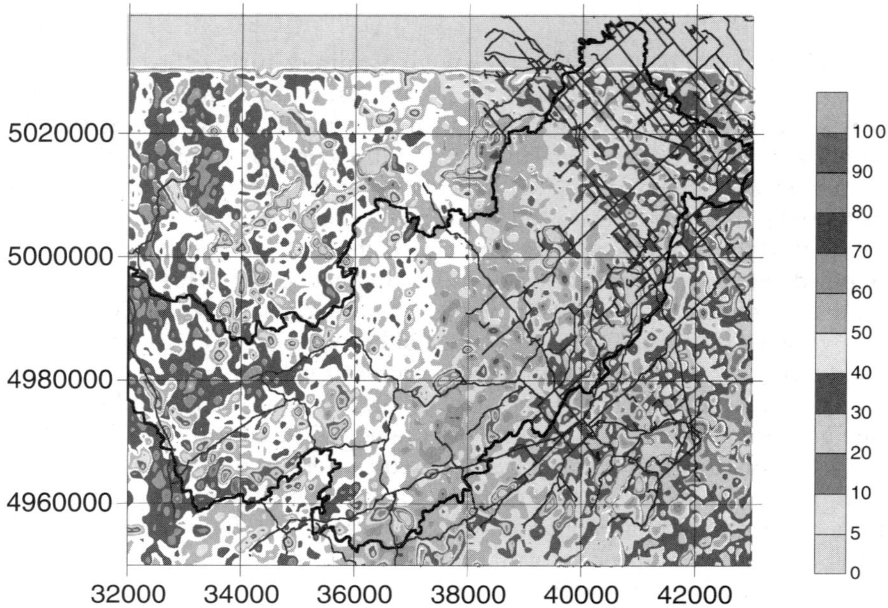


***Figure CS4.5*** RADARSAT-1-derived soil moisture map (in percent) for the Mississippi River Valley Watershed, Ontario, Canada. (From Kije Sipi Ltd. With permission.)

The main reason for the high computed soil moisture values lies in the systematic application of the derived relation (i.e., defining an empirical relationship between RADARSAT-1 image attributes and measured soil moisture from the ground). The relation was applied to the entire RADARSAT-1 scene, irrespective of the actual land cover type, although it was developed from data for open (not under enclosed forest canopy) areas only. For any given soil moisture condition, the radar backscatter signatures are radically different for each land use. When compared with the API map for the project, an east to west trend was only observed in the eastern portions of the maps.

On the basis of these findings, Jobin[176] concluded that a refined soil moisture–mapping technique could likely be obtained by interpolating soil moisture across the entire study area using derived soil moisture averages in the open areas only. This is akin to using the Thiessen polygon spatial interpolation technique for rain gauge data. The proposed procedure includes the following steps:

- Isolate all open areas of at least 100 pixels in size using LANDSAT.
- Determine average backscatter values in each of these areas.
- Calculate the mean soil moisture and API for each of these areas using the derived appropriate relation.
- Spatially interpolate soil moisture and API for the entire watershed using the open area values.

The resulting RADARSAT-1–derived soil moisture maps from this pilot project were subsequently exported in ASCII format and directly used for calibrating the hydrologic model (i.e., updating the surface moisture state variable).

## *CS4.4 Use of LIDAR data for watershed health monitoring*

### *CS4.4.1 Overview*

Some of the most innovative remote-sensing technologies do not involve images at all. Airborne laser mapping, or LIDAR, is a technology that is used for collecting position and elevation data from a pulse-scanning laser system flying in an aircraft. A LIDAR system consists of three technologies: a laser system used for ranging, combined with an inertial navigation system and a GPS for an accurate definition of the aircraft position. A mirror is used to scan back and forth, directing the laser pulse to collect elevation and land cover data while traversing an area (as shown in Figure CS4.6). The points represent laser pulses that have reflected off of treetops, roads, buildings, and the terrain. The points can then be interpolated using mathematical algorithms common in GIS and image analysis packages to model the surfaces.

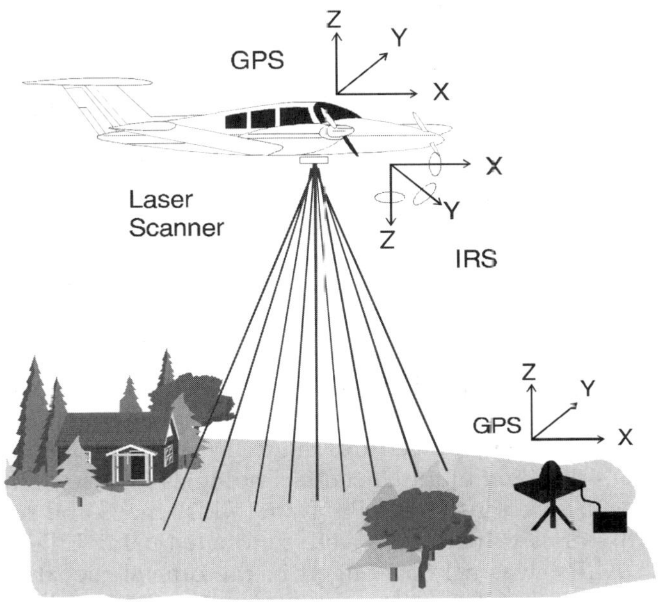


***Figure CS4.6*** Conceptual model showing how LIDAR technology can be used to remotely sense land cover and elevation information using a combination of laser, global positioning system (GPS), and inertial navigation system (INS). (Courtesy of the Foster Creek Demonstration Project "Partners," including Greenland International Consulting Inc., 4DM Inc., Optech Inc., and Ganaraska Region Conservation Authority.)

To extract the terrain, a filter is applied to remove all nonterrain points. The remaining points are then interpolated to produce a digital terrain model (DTM). The DTM from LIDAR data is in WGS-84 horizontal and vertical (ellipsoid height) data. To translate the DTM into a digital elevation model (DEM), a geodetic transformation is required to move from WGS-84 to NAD83 orthometric height. The term *orthometric* is analogous to mean sea level elevation. The data are then run through a rigorous quality assurance and quality control (QA and QC) procedure (as advocated in Chapter 3) to ensure full suitability for contour mapping within the particular specifications of a given project.

LIDAR systems yield elevation data with accuracy that ranges from 0.5 m to less than 1 cm, depending on the altitude of the aircraft and frequency of the readings. One advantage of using LIDAR to acquire elevation data is that it radically reduces the processing time compared with that for other methods currently available. As the data from all of the system components are digital, the LIDAR data service provider can feed them easily into the software processing system and derive usable elevation information in a few days. Other advantages of using LIDAR include the following:

- High vertical precision and high point density.
- Data are in digital format (ASCII xyz) and are multidimensional and directly georeferenced.
- The technology has been thoroughly tested and is commercially available.
- Data can be acquired under cloud conditions or at night.
- No aerotriangulation, orthorectification, or mosaiking is needed.
- Elevations can be obtained for forested areas because the laser scanning system penetrates gaps in the tree canopy.
- Compared with the case of more familiar aerial photographic methods, costs are currently comparable and are fast becoming more economical for large areas (approximately U.S. $2–5 per acre).

In the United States, the Federal Emergency Management Agency (FEMA) has already accepted LIDAR as a tool for flood risk management. Starting in 1997, FEMA began updating and modernizing flood risk mapping and risk assessment procedures in the United States. The initial work included an assessment of advanced technologies for use in the preparation of National Flood Insurance Program (NFIP) maps and related products. In 1998, extensive field tests were conducted using LIDAR, whereby the Optech ALTM was acknowledged in the official report as the "best performing commercial system during the U.S. Government's FEMA tests." Thereafter, in April 1999, an appendix to the U.S. *Flood Insurance Study Guidelines and Specifications for Study Contractors* was released that presented guidelines and specifications for the application of LIDAR to create digital elevation models, digital terrain maps, and other NFIP products.

Some of the emerging applications for LIDAR systems include mapping forested and vegetated areas, as well as surfaces with very little or no texture features, urban areas, and power utility lines (i.e., for maintenance and ice storm contingency planning and operations). In particular, LIDAR can be used for the following types of watershed planning and monitoring projects:

- Developing digital land surface models
- Developing three-dimensional image maps and visualizations
- Assessing stream geometry and geomorphology
- Mapping areas of significant flood risk
- Urban modeling
- Quantifying aggregate extractions from ground quarries, etc.
- Mapping forest or wetland habitat (e.g., tree density, forest patch sizes, etc.)
- Assessing erosional or other landscape changes over time and space

## *CS4.4.2 The Foster Creek LIDAR Pilot Project*

In 1999, Greenland International Consulting Inc. of Ontario was responsible for water management investigations within the Foster Creek system, as part of a subwatershed study by the southern Ontario municipality of Clarington and the Ganaraska Region Conservation Authority (GRCA; the lead watershed agency in the area). Greenland International not only prepared hydrologic and hydraulic computer models of the subwatershed but also was responsible for surveying stream cross sections to support floodplain mapping and fluvial geomorphologic study components. The auxiliary surveying was also intended to check the accuracy of the 1970s base mapping that is almost exclusively applied within the area as a platform for all geospatial analysis. Because Greenland International's fieldwork revealed that the 1970s base mapping was not representative of the subwatershed topography (a similar conclusion was reached later by the GRCA for other watersheds within its jurisdiction), Greenland International and 4DM Inc., Toronto, Ontario, introduced the GRCA to LIDAR remote-sensing technology for future watershed planning, water resources modeling, watershed monitoring, and land use policy development.

In September 2000, the GRCA, in partnership with the Municipality of Clarington and Ontario Ministry of Natural Resources, initiated an innovative use of high-resolution remote-sensing technology. The pilot project was undertaken for the Foster Creek Watershed (east of Toronto, Ontario), as shown in Figure CS4.7. The project was undertaken to assess subwatershed planning techniques, environmental monitoring, and integration with hydrotechnical computer programs, such as flood-line mapping models. Subsequently, the information was used to identify other long-term uses (e.g., municipal and land use planning, utility operations and management, insurance industry needs, master infrastructure design, and planning projects, etc.) and to contrast costs and benefits to those of conventional baseline mapping and data collection techniques. The project managers were 4DM Inc., and the project team

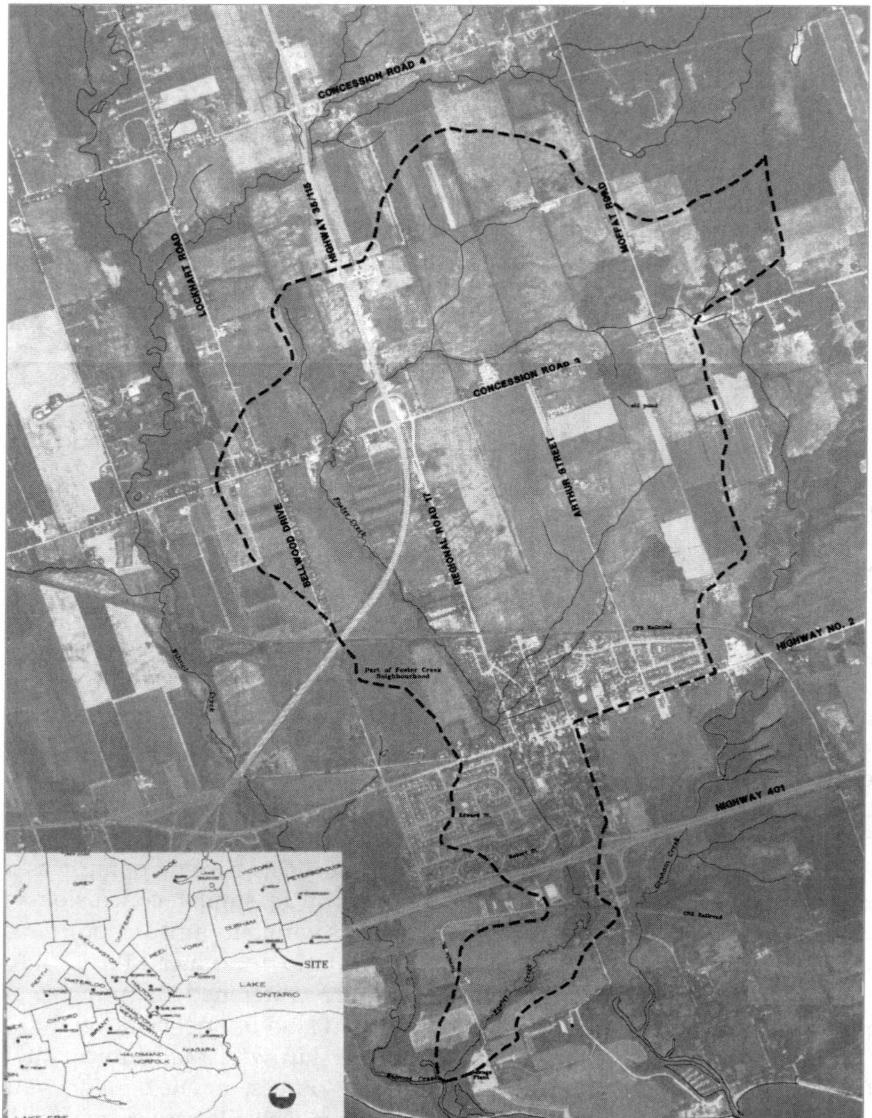


*Figure CS4.7* Location of the Foster Creek Watershed, Ontario, Canada. (Courtesy of Greenland International Consulting Inc. and Gartner Lee Ltd.)

also included Greenland International; Optech Inc., Guelph, Ontario; and Sustainable Development and Monitoring Inc., Waterloo, Ontario.

This pilot project compared the LIDAR mapping data with field-surveyed cross sections of the floodplain from the previous investigations of the Foster Creek system. Another project phase used the LIDAR data for a new HEC-RAS computer model of Foster Creek for floodplain management and flood-forecasting purposes. Greenland International assessed cost-effective field verification methods for this new approach to flood-line mapping in Ontario. The application of LIDAR as an emerging technology for watershed health monitoring of the Foster Creek system also strengthened political linkages (Chapter 2) and community education and involvement (Chapter 4) linkages. For example, project team members were invited to participate on federal and provincial committees examining the widespread use of LIDAR technology for water management across Canada.

#### *CS4.4.2.1 Methodology*

The LIDAR data for the Foster Creek pilot project area (~10 km$^2$ or 2471 acres) was acquired by the ALTM 1225, manufactured by Optech Inc. The remotely sensed elevation data were then used to generate a DTM of the area, which was field-checked for QA and QC purposes. Image analysis and GIS software tools by 4DM Inc. were also used to generate the deliverables.

#### *CS4.4.2.2 Conclusions*

Figure CS4.8 illustrates the high degree of vertical precision inherent in LIDAR data sets as evident from partial coverage from the Foster Creek pilot project. The coverage actually is sufficient to see hydrotransmission lines and waves on Lake Ontario.

***Figure CS4.8*** IKONOS and LIDAR image of the Foster Creek Watershed and Lake Ontario Shoreline. (Courtesy of the Foster Creek LIDAR Demonstration Project partners, including Greenland International Consulting Inc., 4DM Inc., Optech Inc., and Ganaraska Region Conservation Authority.)

*Figure CS4.9* IKONOS and LIDAR image of a watershed in the United States. (From Space Imaging Corporation. With permission.)

A similar image for a river basin in the United States is given in Figure CS4.9.

The project team concluded that LIDAR remote sensing is a valuable surveillance tool for applications in which a high level of land cover precision is required (particularly in cases where applications need precise elevation information, as is the case in flood-forecasting studies). The technology is transferable across a number of watershed health monitoring applications anywhere in the world.

One of the major drawbacks to LIDAR, however, is the cost associated with data collection. Our experience has been that a closed-loop approach with a strong cost recovery partnering model (see Chapter 5) can make LIDAR feasible for at least medium-sized subwatershed projects. The Foster Creek project used a *fracturing cost model* to increase the feasibility of data acquisition using LIDAR technology. The model managed contributions from multiple partners that included municipal governments, utility companies, and others. The fracturing-cost model meant that the project was managed much as in corporate jet ownership plans. The model allowed purchase of a fixed number of dedicated hours per year on a LIDAR-equipped aircraft, based on the fraction of the equipment owned by the project team. The approach reduced the capital expenditure required, optimized use of the instrument by removing downtime, accounted for depreciation of assets, and provided a degree of flexibility to customize flight packages according to project needs and budgets.

## CS4.5 Case study project partners

Participants in this case study included Kije Sipi Ltd, Ottawa, Ontario, Canada; 4DM Inc., Toronto, Ontario, Canada; Optech Inc., Guelph, Ontario, Canada; Space Imaging Corporation, Thornton, CO, U.S.A.; Greenland International Consulting Inc., Concord, Ontario, Canada; and Ganaraska Region Conservation Authority, Port Hope, Ontario.

# *Case Study 5*

# *The Lake Simcoe environmental management strategy*

In Chapter 1, we acknowledged that watershed management is an adaptive process. Watersheds are always in a constant state of change as a consequence of natural processes and human activities. Environmental monitoring techniques provide resource managers with the means to collect and analyze the effects of this change, identify problems, establish targets, develop management strategies, and (based on their effectiveness) make modifications to improve them. It is only through continued environmental monitoring that resource managers can make informed decisions on the necessity of adjusting programs and policies to achieve resource targets and address new or emerging problems.

This adaptive approach to watershed management has been practiced throughout the implementation of the Lake Simcoe Environmental Management Strategy (LSEMS), an ongoing campaign to restore a self-sustaining cold water fishery in Lake Simcoe. The cold-water fish community was selected as the focus of the goal statement because it is an excellent indicator of water quality and ecosystem health. A major objective of the LSEMS program is to reduce phosphorus loading to the lake in an attempt to increase and maintain dissolved oxygen (DO) concentrations. Depressed end-of-summer oxygen concentrations in the bottom waters of Lake Simcoe were considered to be the most significant factor limiting the recruitment of cold-water fish such as lake trout and lake whitefish.[179,180] These species are at present maintained in the lake through stocking efforts and could eventually disappear if efforts to reduce the phosphorus entering the lake are not successful.

LSEMS is a partnership between the Province of Ontario, regional and local municipalities, the Lake Simcoe Region Conservation Authority, and the watershed community. As such, the activities described represent the efforts of many agencies and individuals who are working together to protect and restore the health of Lake Simcoe and its watershed for future generations. The purpose of the this case study is to examine and evaluate the LSEMS program relative to the Closed-Loop Model (Figure 1.3) through a review of its four key components: political support; scientific assessment; community education and awareness; and cost recovery. Discussions will center on the successes and failures of LSEMS to provide practitioners with a better understanding of the watershed health monitoring model and to assist them in avoiding some of the pitfalls that can invariably influence the sustainability of a long-term watershed management program.

## *CS5.1 Background*

Located in southern Ontario, Lake Simcoe is a little more than 1 hour's drive from half the population in the province (see Figure CS5.1).

The watershed has a total land and water surface area of 3534 km$^2$, of which the lake occupies about 20%, or 720 km$^2$. Lake Simcoe is Ontario's largest body of water next to the Great Lakes and is part of the Trent-Severn

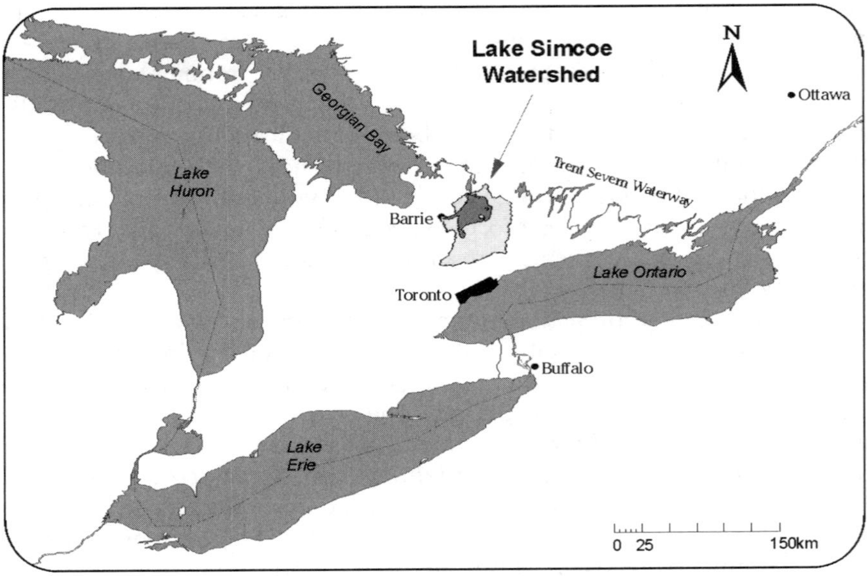


***Figure CS5.1*** Location of the Lake Simcoe Watershed in southern Ontario. (From Lake Simcoe Region Conservation Authority. With permission.)

Canal System, a 386 km navigable waterway connecting Lake Ontario to Georgian Bay. The watershed contains 35 tributaries with five major tributaries draining more than 58% of the total area (see Figure CS5.2). Most of these rivers originate along the southern boundary of the watershed in the prominent glacial feature known as the Oak Ridges Moraine.

For centuries, Lake Simcoe has served as a valuable natural heritage and recreational resource. Fishing, boating, and other recreational activities around the lake are estimated to contribute more than CDN $200 million annually to the local economy. The lake also provides drinking water for five communities and is used to assimilate municipal wastewater discharged from seven sewage treatment facilities. The watershed currently supports a population of approximately 300,000 people, and this number is expected to almost double over the next 20 years.

## CS5.2 Political support

Our Closed-Loop Model maintains that political support is vital to ensure the success of any environmental management program. As we discussed in Chapter 2, political support can ensure that programs receive suitable resources, that activities are coordinated at the appropriate level of government, and that recommendations involving the development of legislation or policy are acted on. In turn, resource managers have an obligation to inform politicians of environmental problems and involve them in the development and implementation of solutions to address these problems.

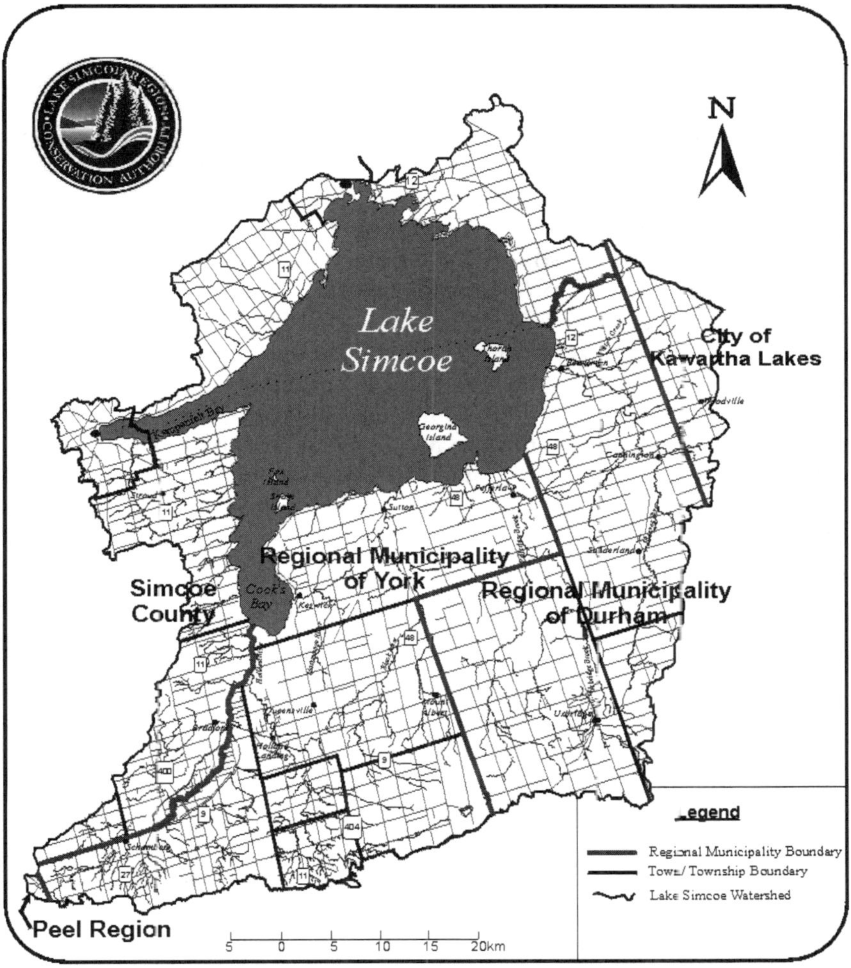


***Figure CS5.2*** Lake Simcoe Watershed. (From Lake Simcoe Region Conservation Authority. With permission.)

The LSEMS program owes its origin and continued operation to the political support that it has received over its 30-odd years of history. The first instance in which politics played a significant role in the LSEMS program occurred in 1975. Prompted by a growing concern from the community and recent data provided by Ontario's provincial ministries that Lake Simcoe was experiencing environmental problems, local municipal governments began pressuring the province to take action. One local politician in particular, Councilor Shillington from the Town of Georgina, began to champion the cause to restore the health of Lake Simcoe. To obtain support, Councilor Shillington organized a 1-day conference in September 1975 in the lakeshore community of Keswick. The purpose of the conference was to inform stakeholders of the health problems plaguing Lake Simcoe and build support to influence all levels of government to take action.

The Keswick Conference was a huge success and resulted in the formation of two committees, the Lakes Simcoe–Couchiching Report Committee and the Lakes Simcoe–Couchiching Steering Committee. The committees were composed of representation from all three levels of government (federal, provincial, and municipal). The Report Committee was directed by the Cabinet Committee on Resource Development to assess the types and magnitudes of environmental problems in the Lake Simcoe–Couchiching watershed; identify the causes of the problems; and propose a strategy for dealing with them. The resulting report, entitled *Lake Simcoe–Couchiching Environmental Strategy,* was released in 1979 and led to the immediate upgrading of sewage treatment plants (STPs) within the watershed to improve phosphorus control. It also was responsible for the launch of the Lake Simcoe Environmental Management studies in 1980 to identify and quantify rural nonpoint sources of phosphorus pollution and recommend remedial measures and control options to achieve an additional 4-tonne (metric tons) reduction in phosphorus per year.

It is extremely important, once political support is obtained, that the momentum be maintained. On completion of the LSEMS studies in 1985,[194] significant reductions in phosphorus loading to Lake Simcoe had been achieved through point source controls (upgrades to the STPs and a diversion of sewage from the communities of Newmarket and Aurora outside of the watershed). These activities had been estimated to reduce phosphorus loads entering the lake by 8 tonnes/year; however, the improvements had come at a high cost (more than CDN $130 million) to both the province and municipal governments. A provincial cabinet submission requesting further provincial investment to address agricultural sources of phosphorus and recommending the adoption of phosphorus control regulations was stalled for 4 years before the cabinet granted approval. Although there is no clear understanding of why the approval process took so long, several factors may have contributed to the delay.

First of all, the key individuals or champions (i.e., Councilor Shillington) responsible for past successes in moving the program forward were no longer involved in the political process. At the same time, political pressure exerted by the community had been reduced. Seeing that action was being taken (STP upgrades), the community became confident that environmental problems were being addressed and relaxed its efforts. Furthermore, catches of cold-water sport fish were on the increase and could have been interpreted by the public as proof that the health of the lake was improving (in reality, the improved catches were not associated with increased recruitment but were a result of stocking efforts of the Ontario Ministry of Natural Resources). The cabinet submission also got sidetracked because of two provincial elections occurring in 1986 and 1989. From a technical standpoint, the additional phosphorus target established in 1979 (4 tonnes/year) was somewhat arbitrary. A comprehensive relationship between phosphorus loading and DO had not been established, so the severity of the problem was not apparent. Finally, communication efforts associated with the LSEMS program were discontinued in 1985 with the conclusion of the LSEMS studies.

Despite the lengthy delay, the cabinet submission eventually received approval, and in 1990, the local Members of Parliament announced the first phase of the LSEMS implementation program. The cabinet submission directed that the Lake Simcoe Region Conservation Authority (LSRCA, the lead watershed agency) take the lead role in implementation and report back in 1995 regarding the progress attained. Fortunately, monitoring programs instituted as part of the LSEMS studies continued in the intervening time because of internal support from district offices of the Ontario Ministries of Environment and Natural Resources. This ensured completeness of the data sets, enabling resource staff to track long-term trends over time.

After 5 years of successful implementation, the completion of more than 300 remedial projects, and an estimated reduction of 7.5 tonnes/year of phosphorus from agricultural sources, lack of political support plagued the LSEMS program again. In 1996, a second cabinet submission was being prepared for a second phase of LSEMS. The need for a second phase had been determined through monitoring data used to develop an understanding of the relationship between phosphorus loading and DO. As a result, a new phosphorus target was established, recommending that the annual phosphorus loading entering Lake Simcoe should not exceed 75 tonnes/year. Current estimates were that the average annual phosphorus load entering Lake Simcoe was about 100 tonnes, suggesting that an additional 25 tonnes of phosphorus must be reduced to achieve the LSEMS goal of a self-sustaining coldwater fishery. Just as the cabinet submission was being prepared, the provincial government changed once again, and shortly thereafter, all provincial capital funding to Conservation Authorities across Ontario was discontinued as a result of provincial restructuring and cost-cutting measures. This meant that all of the provincial LSEMS capital funding to undertake remedial projects was lost (approximately $350,000 annually); however, as in 1985, funding for continued monitoring was maintained through the regional provincial offices.

With this loss of a primary funding partner, the Authority looked to other partners and new fee-for-service charges to boost revenues. One of the fundamental strengths of Conservation Authorities in Ontario is that they are governed by a board of directors comprising members appointed by local governments within the watershed area. Member municipalities supported the Authority's efforts, and between 1996 and 2000, the LSEMS Implementation Program grew despite provincial cutbacks. Achievements of Phase II were noticeably less than those of Phase I, with only 55 projects completed, accounting for a modest reduction of 2.5 tonnes/year; however, the lessons learned regarding the need for continued political support at all levels of government were not lost on the LSEMS partners.

At present, the Authority, as the lead agency of the LSEMS Implementation Program, is negotiating with its member municipalities and the Province of Ontario for a third phase. A series of public and mayoral meetings have been organized to build municipal support and exert pressure on the province to reinvest in LSEMS. Appropriately, the initial public and mayors meetings were held in the Town of Georgina, where the very first meetings were held to develop the original LSEMS initiative to protect Lake Simcoe. Meetings have also been held to inform federal Members of Parliament (MPs) and Members of Provincial Parliament (MPPs) of the severity of the environmental problems facing Lake Simcoe, as well as the necessary remedial actions needed to alleviate these problems. Other factors contributing to political support for LSEMS are discussed below. First, the creation of a new political committee as part of the LSEMS Phase III governance model formalized political linkages. All other committees report to the political committee, which comprises the LSRCA's board of directors, mayors, regional chairs, county wardens, MPs, and MPPs, to ensure that there is communication between all levels of government.

Development of business and work plans for LSEMS Phase III every 3 years to correspond with the municipal government terms of office assisted in setting milestones that worked in the political forum of local government. The Phase III Memorandum of Understanding will be signed by the provincial agencies, municipalities, and the Conservation Authority and will remain in place for 6 years to ensure that provincial support is maintained between the electoral terms of office. Furthermore, education and awareness programs that are intrinsically linked to political support were enhanced.

The LSEMS experience supports the inclusion of political linkages in the Closed-Loop Model. Resource agencies can only be successful if appropriate resources are available. We hope that the actions taken above will provide politicians with the facts necessary to make informed decisions and allow them to share in the responsibility for the protection of Lake Simcoe.

## CS5.3 Scientific assessment

Chapter 3 discussed scientific assessment methods for evaluating watershed health and stated that distinguishing between impaired and unimpaired conditions and determining causation in the aquatic environment frequently require collection of both biotic and water chemistry information. The LSEMS studies and previous investigations that were conducted to understand why the cold-water fishery in Lake Simcoe was declining embraced this approach. The Ontario Water Resource Commission (at present the Ontario Ministry of the Environment) conducted the first comprehensive studies of the water quality of Lake Simcoe between 1971 and 1974. Additional studies were completed between 1976 and 1979 and again between 1980 and 1985. Since that time, the Ministry of the Environment and the LSRCA have continued to monitor water chemistry in the lake and its tributaries. In 1960, the Ministry of Natural Resources established the Lake Simcoe Fisheries Assessment Unit (LSFAU) to monitor trends in the fish community of Lake Simcoe and to evaluate the effects of factors such as fishing, habitat degradation, invading species, weather patterns, and changes in water chemistry. The data collected have been vital to the identification of water quality problems; establishment of a phosphorus-loading objective for Lake Simcoe; development of a phosphorus control management strategy; and the evaluation of spatial and temporal water quality trends as a means of performance evaluation relative to the phosphorus objective. A description of the methods used to collect data and integrate monitoring information into the Lake Simcoe management cycle follows.

## *CS5.3.1 Open-lake monitoring*

A total of 12 sampling stations (as illustrated in Figure CS5.3) are visited about nine times a year between ice-out in spring (usually between April and May) and fall turnover (usually late October).

Specific monitoring activities consist of the following:

- Measuring temperature and DO at 1-m intervals within the water column using an oxygen temperature probe (Yellow Springs Instrument 58, Yellow Springs, Ohio, U.S.A.) and Secchi disk measurements for visibility
- Collecting composite water samples of the euphotic zone for analysis of inorganic and total forms of nitrogen, phosphorus, silica as well as chlorophyll *a* and phytoplankton

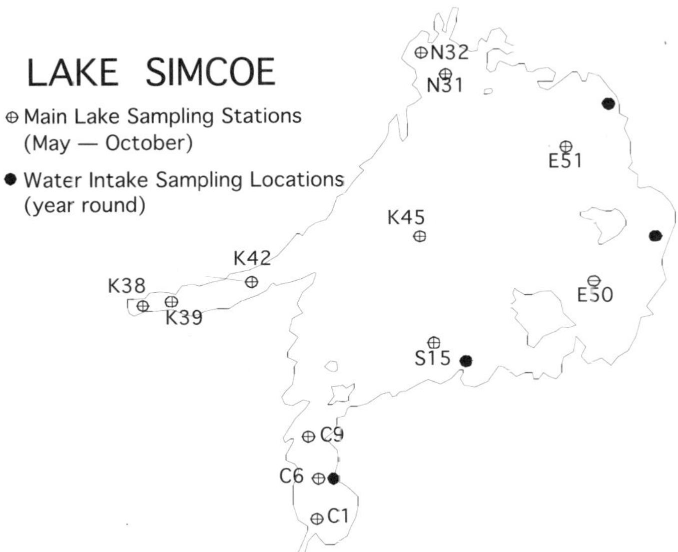


***Figure CS5.3*** Open-lake sampling stations. (From Nicholls, K.H., 1998 Lake Simcoe Water Quality Update with Emphasis on Phosphorus Trends, LEMS Report Imp.B.17., Lake Simcoe Region Conservation Authority, Newmarket, Ontario, 1998. With permission.)

- Collecting additional samples at 1, 5, 10, and 15 m above the lake bottom at deeper stations to be analyzed for nutrients and some metals

In addition to the open-lake stations, the Ministry of the Environment collects weekly samples of raw, or untreated, lake water with the assistance of the municipalities taking water from the lake. The water is taken to be treated and provided as potable water for five communities. This is a cost-effective way to collect samples during times when it is too dangerous to venture out on the lake (i.e., ice-out conditions). The open-lake water quality data were collected to identify why the cold-water fishery population was declining, develop a phosphorus-loading objective, assess the effectiveness of remedial measures and control options over time, and identify any new or emerging problems.

### CS5.3.2 Tributary monitoring

The tributary monitoring program was initiated to collect baseline chemical and physical water quality and stream flow quantity data from 13 sites along four of the five major tributaries draining more than 50% of the watershed area (see Figure CS5.4).

Limited resources and the large number of smaller watercourses in the northern watershed made it impractical to monitor all the tributaries entering Lake Simcoe. The tributary monitoring program was developed to assist in estimating the total phosphorus load entering Lake Simcoe; qualify the distribution of phosphorus sources within the watershed; calibrate a predictive computer model; assess the effectiveness of remedial measures using trends over time; and identify new or emerging problems.

Sites are sampled weekly and more frequently on an event basis using ISCO™ 2150 automatic samplers and grab samples where rainfall or snowmelt resulted in increased stream flow. Two of these were located at pumping stations that regulate water levels within the Holland Marsh, a 27-km$^2$ area of drained wetland used extensively for market gardening and vegetable production. Pumps were calibrated so that water volumes could be estimated based on pumping time. Automatic samplers were installed to collect composite samples while the pumps were operational, thereby providing the chemical water quality concentrations needed to calculate annual loads. Analytical parameters tested included suspended solids, total and filtered reactive phosphorus, total Kjeldahl nitrogen, total nitrogen, total ammonia, and chloride.

Water quantity was estimated using Stevens water level recorders and through the development of stage-discharge rating curves. This method allowed a continuous estimate of stream flow in cubic meters per day. Tributary monitoring stations were located near the mouths of most tributaries to provide a measurement of as much of the subwatershed area as possible. The term *loading* is used to describe the total amount of a pollutant measured entering or leaving a system. Loads are calculated by multiplying concen-

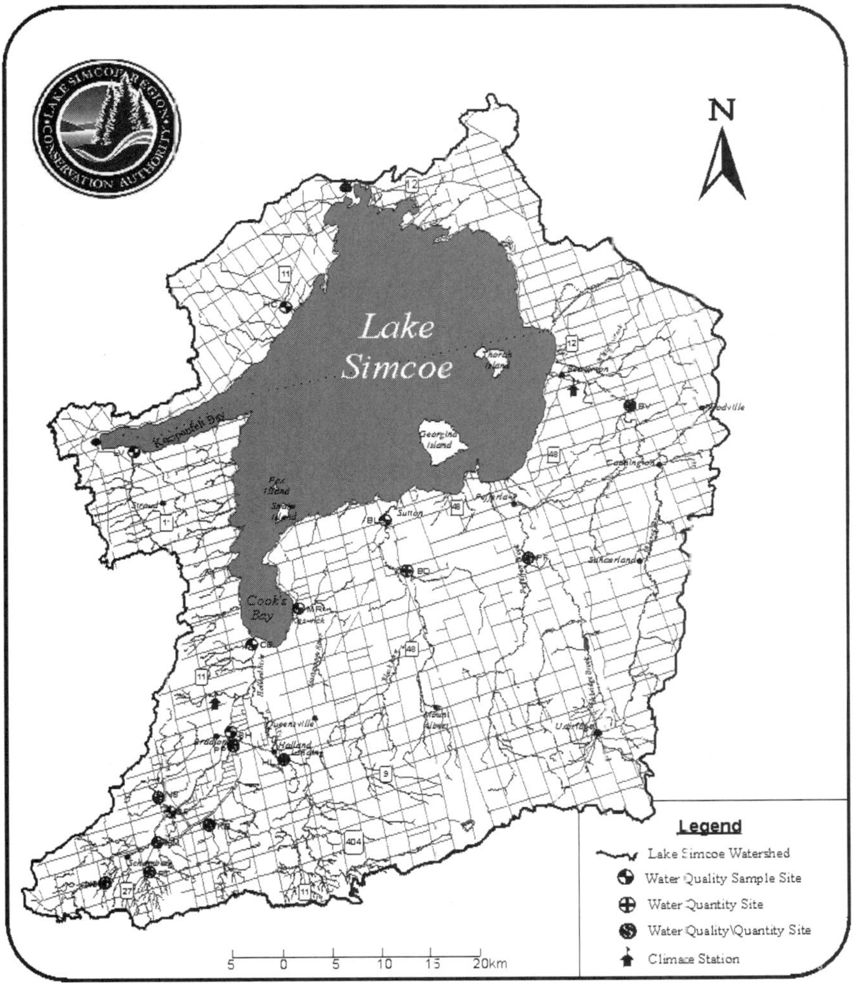


*Figure CS5.4* Tributary monitoring stations in the Lake Simcoe Watershed. (From Lake Simcoe Region Conservation Authority. With permission.)

tration times the flow volume and are expressed in units of mass per unit time. Two methods have been used, the *Beale Ratio Estimator*[177] and the *Around Method*.[178] Loads are generally reported in tonnes per year but have been reported in kilograms per day or month when used to depict seasonal trends or document a significant storm or snowmelt event.

As mentioned, more than 65 individual watercourses drain to Lake Simcoe, but only 13 tributary sampling sites could be established because of limited resources. To ensure that loading contribution from the unmonitored portions of the watershed was assessed, unit–area loadings estimates for monitored subwatersheds containing similar physical and land use characteristics were calculated and applied to the unmonitored areas. A unit–area load is estimated by simply taking the total (generally annual) load for a monitored subwatershed and dividing it by the total area of the monitored subwatershed. Annual unit–area loads in the Lake Simcoe watershed are represented in kilograms per square kilometer or hectare.

One of the main benefits of using loading estimates is that the data are easy to interpret, especially by the general public. The visual impact of graphing the relative loading contribution from each subwatershed immediately provides the average person with a clear understanding of where the phosphorus is originating within the Lake Simcoe watershed. This allows resource managers to target their programs for phosphorus control. Unfortunately, the loading estimates produced from the monitoring did not have reach-level resolution and could not convey exactly where within subwatersheds the phosphorus was arising or the relative importance of individual source contributions (at the site level). Even at a relatively coarse subwatershed level, however, this information was the key to developing a meaningful management strategy, one that would ensure that remedial measures targeted the most significant contributors.

### *CS5.3.3 Lake Simcoe Fisheries Assessment Unit*

The LSFAU monitors the status of cold-water fish in Lake Simcoe by conducting index netting programs and creel surveys of the winter fishery.[179] A creel survey involves interviewing anglers while they are fishing to determine how many fish of various species they have caught and how long it took to catch them. Catch records are also compiled and provided to the LSFAU by fishing charter operators and anglers. The size and weight of some harvested fish are also measured. These data provide estimates of the total number of fish caught, trends in the catch, the amount of fishing pressure that is occurring, and the condition of fish caught. The LSFAU is also responsible for collecting the eggs and spawn from lake trout and lake whitefish to be used in restocking efforts that are currently maintaining the cold-water fishery in Lake Simcoe.

## *CS5.4 Defining the problem*

Conclusions drawn from investigations into the water quality of Lake Simcoe indicated that although the general water quality was adequate for most of the PWQO,[75] the lake's trophic (or "nutrient enrichment") state was out of balance because of human land use activities associated with urbanization and agriculture. Poor farming practices and urban development were responsible for accelerating the release of nutrients, in particular phosphorus, into Lake Simcoe. The additional phosphorus was responsible for an increase in aquatic plant and algal growth, as well as a decline in water clarity. In addition to the aesthetic impairment associated with the algae and aquatic plant growth, decomposition in the bottom waters of Lake Simcoe was resulting in depressed levels of DO.

Investigations into the health of the cold-water fishery found that although invading species, habitat degradation, and fishing pressure were affecting populations, the loss of DO was the major factor responsible for the decline in the cold-water fish population.[180,181] Each year, before fall mixing of the lake, the concentration of DO in the bottom waters was observed to drop to between 1 and 3 mg/l, well below the minimum level of 4 mg/l required by lake trout and lake whitefish. Most of the spawning shoals were located in the areas of oxygen depletion, and it was concluded that the cold-water fish species were not able to reproduce in the oxygen-poor environment and, as a result, were declining in number. It is on these two premises, (1) that phosphorus loadings influence oxygen deficits and (2) that oxygen deficits are responsible for the decline in the cold-water fish populations, that the entire LSEMS implementation program is based.

## *CS5.4.1 Developing a phosphorus-loading objective*

With the problem identified, an understanding of the relationship between phosphorus loading and DO depletion was needed to set a phosphorus objective for Lake Simcoe. To this end, two independent models were used to explain the phosphorus–DO dynamics within the lake. Snodgrass and Holubeshen[182] developed the first model in 1993. It was a probabilistic simulation based on the mechanics of the growth and decay of organic matter relative to phosphorus loading. The deterministic model was based on a chemical engineering concept of a "stirred tank reactor" and involved multiple Monte Carlo simulations, whereby probability distributions rather than single values were specified for input parameters. Subsequently, results of the model were presented as frequency distributions. The model was originally run on the assumption that the annual phosphorus load entering Lake Simcoe averaged 66 tonnes/year. This total load was later estimated to be about 100 tonnes/year,[183,184] and as a result, the *x* axis or phosphorus-loading rate was rescaled to reflect the new average annual load.

The second model was developed in 1995 by Nicholls,[135] who reasoned that if two distinctly different models predicted similar responses, then confidence in the model predictions would be increased. Nicholls used an empirical modeling approach relating DO concentration to phosphorus loading through ordering intermediary relationships among trophic-state variables (see Figure CS5.5). This method used open-lake monitoring data for DO, chlorophyll *a*, and total phosphorus concentrations. Similar relationships have been evaluated with empirical data sets from around the world.[186–192]

A comparison of the results of the two models was extremely encouraging as predictions from both models showing excellent agreement. The revised Snodgrass–Holubshen model predicted that DO concentrations of 4, 5, and 6 mg/l could be achieved by reducing the annual phosphorus-loading

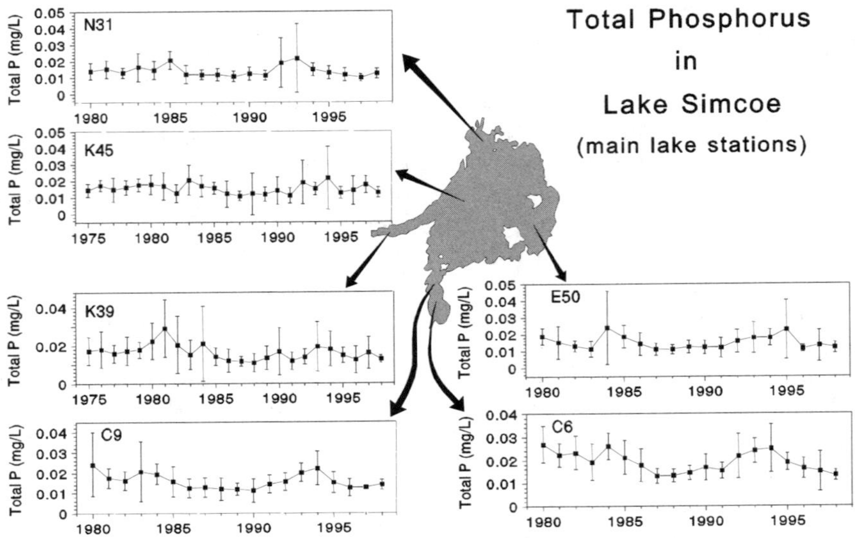


*Figure CS5.5* Total phosphorus trends in the main lake stations. (From Nicholls, K.H., 1998 Lake Simcoe Water Quality Update with Emphasis on Phosphorus Trends, LEMS Report Imp.B.18, Lake Simcoe Region Conservation Authority, Newmarket, Ontario, 1998. With permission.)

rate from the present-day average of 100 tonnes/year to 89, 72, and 55 tonnes/year, respectively. The Nicholls model predicted the same results within 1%. Nicholls then set about developing equations for predicting DO concentrations at different total-phosphorus loads and phosphorus-loading rates that achieve given DO concentration. By plotting the two predictive curves and defining the predicted end-of-summer DO concentration as the midpoint between these curves, Equation CS5.1 applies for the relationship between DO and total phosphorus in Lake Simcoe:[185]

$$DO = 10.73 - 0.094 \times P_L + 0.0002 \times P_L^2 \qquad \text{(CS5.1)}$$

where DO is the mid-September DO concentration in milligrams per liter, volume weighted through the 18 m to bottom zone, and $P_L$ is the phosphorus-loading rate to Lake Simcoe in tonnes per year.

Alternatively, the equation for phosphorus loading (the converse relationship between total phosphorus and DO in Lake Simcoe) is listed below in Equation CS5.2:[185]

$$P_L = 229 - 69.9 \times DO^{1/2} \qquad \text{(CS5.2)}$$

where DO is the mid-September DO concentration in milligrams per liter, volume weighted through the 18 m to bottom zone and $P_L$ is the phosphorus-loading rate to Lake Simcoe in tonnes per year.

With the phosphorus–oxygen dynamics established, the next step in the determination of the phosphorus-loading objective was to select a suitable end-of-summer DO concentration that would satisfy the requirements of the cold-water fishery. To this end, efforts to determine the physiological response of lake trout to different levels of oxygen were initiated in a laboratory, and the DO concentrations in other Ontario lakes known to support healthy lake trout populations were examined.[181] An end-of-summer DO concentration of 8.0 mg/l was considered optimal for lake trout; however, the corresponding loading rate in Lake Simcoe to achieve this was approximately 32 tonnes/year, which would require a phosphorus-loading reduction of almost 70 tonnes/year. Given the existing extent of urbanization and agricultural land use within the watershed, a reduction of this magnitude was not realistic. Eventually, an interim end-of-summer DO concentration of 5 mg/l was agreed on, and although it was recognized that lake trout would still experience some stress, the 5 mg/l DO concentration did represent a significant improvement from the previous concentrations of ~3 mg/l. With an interim end-of-summer DO concentration set, the Lake Simcoe phosphorus-loading objective was calculated at 75 tonnes/year, which, based on an average present-day loading of 100 tonnes/year, would require a reduction of the existing phosphorus loading by 25 tonnes/year.

## *CS5.5 Producing a management strategy*

At the time that the phosphorus-loading objective was established, the first phase of the LSEMS Implementation Program was almost at an end. Although a significant reduction in phosphorus loading to Lake Simcoe had been accomplished, much more remained to be done to achieve and maintain an annual phosphorus-loading rate of 75 tonnes/year. An initial requirement of Phase I was that a report be provided to the Cabinet to summarize the progress achieved and recommend further action if necessary. The newly developed phosphorus objective definitely sanctioned further action. To obtain Cabinet support, the Steering Committee requested that a detailed management strategy be developed including recommendations for remedial measures, and regulations to achieve the LSEMS goal. To assist in the development of a comprehensive management strategy, and to ensure that the phosphorus objective could be achieved and maintained, a predictive water quality model was developed by BEAK Consultants Ltd in 1995.[183,184] Its purpose was to account for all the major point and nonpoint sources of phosphorus within the watershed; predict the effect that future urban growth would have within the watershed; and evaluate the effectiveness of various remedial measures and control options to reduce phosphorus loading.

The Lake Simcoe Watershed Model is based on Hydrology-Based Simulation Tool (HYDROSIM) and was adapted from the U.S Environmental Protection Agency (EPA) model HSP-F.[193] The model uses a lumped-parameter approach requiring that the watershed area be discretized into segments with like hydrologic characteristics. A geographic information system (GIS)

***Table CS5.1*** Urban and Rural Sources of Phosphorus Considered Using the Lake Simcoe Water Quality Model

| Urban sources of phosphorus | Rural sources of phosphorus |
|---|---|
| Dry weather discharges from storm sewers | Atmospheric deposition |
| Wet weather discharges from storm sewers | Private waste management systems (septic systems) |
| Sewage treatment plants | Runoff from livestock operations |
| Private waste management systems (septic systems) | Erosion from agricultural lands in production (tilled and pasture) |
| | Diffuse seepage from natural areas (forested lands, wetlands, scrubland) |

*Source:* BEAK Consultants, 1995. Development and Implementation of Phosphorus Loading Watershed Management Model for Lake Simcoe, LSEMS Implementation and Technical Report A.3, Lake Simcoe Region Conservation Authority, Newmarket, Ontario.

was used to overlay the various data layers (soils, land use, digital elevation model, and vegetation) to create homogeneous polygons that were linked to the appropriate subcatchments within the watershed. Weather variables are inputted into the model on a time-step basis (i.e., hours or days) to allow the model to predict outflows to the tributaries and eventually Lake Simcoe. Water quality parameters are then appended to the hydrologic simulation and in-stream routing components, resulting in the estimation of phosphorus loadings. Phosphorus sources considered in the model were divided into urban and rural categories, as listed in Table CS5.1.

A critical step in any model development is calibration. The Lake Simcoe Water Quality model (HYDROSIM) involves the simulation of numerous natural processes based on a series of mathematical equations and process variables. These process variables were determined during the calibration process so that model predictions corresponded to observed data. Stream flow monitoring data for six sites on the five main tributaries, and total phosphorus concentrations at three of the sites, were used to estimate daily loading rates of phosphorus for a 6-year period from 1986 to 1991. Using this method, predicted daily, monthly, and annual phosphorus loads were compared to observed values. Sensitive parameters were developed and adjusted until an acceptable level of accuracy was attained. In general, the average measured stream flow rates and total phosphorus loads corresponded well to the predicted values (less than 1% for stream flow and within 15% for the total annual phosphorus loads).

The result was a powerful predictive tool that could be used to evaluate the change in total annual phosphorus loading to Lake Simcoe associated with what-if or gaming scenarios. Numerous model scenarios were run to determine an appropriate management strategy to achieve the Lake Simcoe phosphorus-loading objective of 75 tonnes/year. In addition, future growth scenarios were also examined to determine whether the objective could be maintained despite additional inputs of phosphorus associated with future urban expansion. Modeling results indicated that the phosphorus-loading objective could be achieved and maintained to the year 2011 based on current growth rates within the watershed. However, after 2011, as urban growth within the watershed continues, phosphorus loads will begin to increase unless new, more effective best management practices and control options can be implemented or plans for further development within the watershed are suspended.

As a result of the modeling exercise, recommendations to implement a full range of urban and rural remedial measures and control options were proposed. Efforts to control existing urban loadings from uncontrolled urban areas were needed. In addition, a special policy requiring the construction of stormwater management control facilities capable of removing 80% of suspended solids (Ontario Ministry of the Environment Level 1 Stormwater Control) was recommended. Most of the Lake Simcoe watershed (approximately 70% of the area) only required Level 2 Stormwater Control (stormwater facilities capable of removing 70% suspended solids) under provincial policy; however, with recognition that Lake Simcoe was the ultimate receiving water and was already experiencing water quality problems, the most stringent level of water quality control was established throughout the watershed. This recommendation required support from the Ontario Ministries of the Environment and Natural Resources and is the responsibility of the Conservation Authority to implement and enforce.

## *CS5.5.1 Examples of adaptive management in the LSEMS Implementation Program*

Assessing the effectiveness of the LSEMS Implementation Program and adapting the management strategy to improve phosphorus controls to achieve the LSEMS goal of a self-sustaining cold-water fishery was one of the primary purposes of the monitoring programs when they were initiated in 1982. A number of changes to the direction of implementation activities have been made as a result of the monitoring data collected over the years.

During Phase I (1990–1995), emphasis was placed on working with the agricultural community to address phosphorus inputs from poor farming practices. This was based on earlier phosphorus-loading estimates indicating that agricultural sources were the main contributors of phosphorus entering the lake.[194] By the end of Phase I, however, phosphorus loading associated with stormwater runoff was increasing significantly as urban expansion within the watershed accelerated. This redistribution of the loading had been observed as massive increases in sediment and phosphorus occurred at a number of sampling stations downstream of urban centers. This trend was further quantified by the modeling efforts and resulted in the development of a stormwater management program offered by the authority to its municipal partners. As a result, urban stormwater management strategies have been completed for almost all of the urban centers within the watershed, and stormwater management retrofit projects are being completed to reduce phosphorus loadings from uncontrolled urban areas. The policies mentioned earlier, to control urban phosphorus inputs from new development, were also initiated as a result of the monitoring data. Furthermore, educational programs such as the Yellow Fish Road have been developed in partnership with local service clubs and the schools. The program involves school children and interest groups painting yellow fish on roadside curbs adjacent to storm drains and distributing information to educate the community regarding the connectivity of storm drains to watercourses draining to Lake Simcoe. Specific emphasis is placed on changing the public's attitude regarding the application of fertilizers and disposal of hazardous chemicals around the home.

The LSFAU has also adopted an adaptive approach in its decision-making process with respect to the management of the cold-water fishery in Lake Simcoe. One of the first management decisions made when the cold-water populations of lake trout and whitefish began to decline was to introduce restocking to maintain the species despite declining natural recruitment. Further decisions have involved limiting catches of both species and even a total ban in 2001 on the possession of lake herring, another cold-water species in decline. The LSFAU uses the data collected from monitoring, public input, and provincial policy to develop arguments for management options; however, it is the district team that has the final responsibility for management decisions.[179]

Without a doubt, the most significant threat to the LSEMS Implementation Program and lake health was the loss of provincial capital funding that was used to provide incentives for landowners and municipalities to undertake phosphorus control projects within the watershed. For every dollar that the province provided, the Authority would match or more than double the monies through partnerships with municipalities or landowners. During Phase II, the estimated phosphorus reduction was 2.5 tonnes/year, much less than the 7.5 tonnes/year reduced in Phase I. With limited capital dollars, all remedial projects were prioritized based on their cost per kilogram of phosphorus reduction. This prompted a review of all the past spending practices on programs or projects to achieve phosphorus reductions within the watershed. It was observed that a huge investment (approximately CDN $130 million) had been allocated to build new or to upgrade older STPs and construct the York–Durham pipeline that diverts sewage from the Towns of Newmarket, Aurora, and Holland Landing (see Figure CS5.2) outside of the watershed for ultimate delivery to Lake Ontario. These measures achieved a total phosphorus-loading reduction of 7.5 tonnes/year. The investment equated to a cost of CDN $17,333 per kilogram of phosphorus reduced within the watershed and sharply contrasted with the effectiveness of nonpoint source projects such as restricting livestock access to a watercourse, which costs about $100 per kilogram of phosphorus reduced.

The above cost–benefit exercise has led to support for the development of a Total Phosphorus Management Program in the Lake Simcoe watershed by the Conservation Authority, regional municipalities, and the Ministry of the Environment. The program will be based on aquatic effluent trading and will be similar to initiatives undertaken by the U.S. EPA.[195] One difference is the inclusion of urban storm water management as an eligible trading component. The rationale is based on a relaxing of the Level 1 Stormwater Control requirements to the less stringent Level 2 Stormwater Control (where appropriate) and levying a cash-in-lieu investment from the developer to the Authority. These funds will then be reinvested into urban stormwater retrofits to reduce phosphorus loadings from uncontrolled urban areas identified within the watershed. The program is still far from the implementation phase. The future viability of the program depends upon the assumption that Level 1 Stormwater Control is only slightly more effective at phosphorus reduction (less than 10%) than Level 2. However, the construction costs associated with a Level 1 facility are almost double those of a Level 2. By accepting a cash-in-lieu payment, the costs of retrofitting all of the remaining uncontrolled urban area within the watershed could be obtained from new development, resulting in a 16 tonnes/year reduction in phosphorus.[196]

## *CS5.5.2 Assessing the effectiveness of remedial efforts*

An evaluation of data from all three of the related LSEMS monitoring programs (open lake, tributary, and LSFAU) allowed conclusions to be made regarding the effectiveness of remedial efforts in relation to LSEMS goals and objectives. These conclusions are essential to determine the future course of action and adapt management strategies accordingly. Unfortunately, phosphorus concentrations within Lake Simcoe have not improved significantly, but they also have not worsened. Trend analysis of open-lake data employed a robust nonparametric statistical technique developed especially for use with water quality results that are characterized by seasonality and serial dependency.[197] An examination of total phosphorus concentrations at the outflow of the lake generally showed lower concentrations of total phosphorus during the 1990s than were observed in the 1980s.[198] However, the results in all probability are not reflective of the remedial efforts associated with diffuse source controls but of more localized improvements of the Orillia sewage treatment plant and the influence of zebra mussels. Similar patterns in the main lake sampling stations have also emerged; however, the magnitude was not as significant as that seen at the outflow station.

The impact of zebra mussel filtration of the water column was again credited for the slight improvement. From these results, it was concluded that phosphorus-loading reductions associated with the remedial measures and other control options have only maintained the status quo within Lake Simcoe by offsetting new source inputs from continued urban growth within the watershed. Correspondingly, DO concentrations in the deep-water areas of the lake have also remained relatively unchanged, and as a result, the status of the cold-water fishery has also not improved. Recent data collected from the LSFAU indicate that there is still no lake trout reproduction occurring in Lake Simcoe. Disturbingly, a further decline in natural lake whitefish recruitment has also been noted, which is cause for concern (Willox, C., personal communication, 2000). The Lake Simcoe whitefish was designated as a threatened stock by the Committee on the Status of Endangered Wildlife in Canada in 1987.[199]

It is clear from the results that to achieve the LSEMS goals and objectives, a significant effort and investment is required to address the remaining phosphorus sources within the watershed. The level of resources committed to controlling nonpoint sources of phosphorus loading over the previous two phases of LSEMS implementation will not be sufficient to achieve the reduction necessary to get to the 75 tonnes/year phosphorus-loading objective. In addition, the issue of continued urban growth within the watershed must be addressed. The LSEMS partners realize that a holistic approach to water management is required under the next phase of implementation. In addition to the capital projects to reduce phosphorus from nonpoint sources, regulations to address continued urban growth within the watershed will have to be evaluated. Any interference to limit growth has been swiftly condemned in the past because of social and economic considerations. Despite this, growth issues will be part of a future study, entitled *State of the Watershed Report* and proposed for the year 2001, to bring the issues to the attention of the watershed community. Two subsequent exercises are planned, culminating in the development of a Lake Simcoe Water Management Strategy and a Lake Simcoe Watershed Plan.

## CS5.6 Community education and awareness

Chapter 4 discusses the importance of involving the community in the watershed-planning process and implementation activities. Unfortunately, the LSEMS studies conducted between 1982 and 1985 did not involve the community in the development of the management strategy. This lack of public participation is believed to be one of the main contributing factors resulting in the poor political support for the LSEMS program. It also has been attributed to a general lack of awareness of some watershed communities regarding the severity of the lake's environmental problems. As a result, the first phase of the LSEMS implementation program was announced in 1990 without the support of much of the watershed community.

In 1992, the LSEMS partners realized their mistake and initiated efforts to develop a public advisory committee to provide the community the opportunity to become directly involved in LSEMS activities. The Save Our Simcoe (SOS) Committee was formed from volunteers throughout the Lake Simcoe Watershed to help educate the community and raise awareness of the lake's environmental problems by acting as a conduit to the public. The group members also wanted the opportunity to get their "hands dirty" by actually assisting in the cleanup effort and organizing volunteer activities to provide the community an opportunity to get involved. The committee governance structure involved a chairperson, secretary-treasurer, and a representation from the LSEMS Technical Committee. The chairperson was elected from the volunteer members and was responsible for managing meetings, attending speaking engagements, and representing the SOS Committee at LSEMS Steering Committee meetings. The last duty ensured that the SOS Committee had a voice in the decision-making process to direct implementation activities. The technical representative and the secretary-treasurer were appointed from the Conservation Authority, which also provided the Committee with a modest budget as seed funding to initiate its activities. It is important to realize that control of the Committee lay with the membership. Conservation Authority staff attended as nonvoting members and were only there to provide advice and support to the committee.

The SOS Committee began almost immediately to make significant contributions to the LSEMS Implementation Program. Public awareness within the watershed of the lake's problems increased dramatically, and community members were empowered to act locally as well as around their own homes to ensure that they were not contributing to the pollution problem. The creation of "Action Guide: To Save Lake Simcoe" was a significant achievement that provided watershed residents and cottagers with helpful hints and instructions on how they could help solve the pollution problems within their communities and around their homes. In addition, the committee began organizing volunteer days to clean up the watercourse and complete environmental improvement projects (i.e., planting days to establish riparian buffers, stream bank erosion control projects). Partnerships with schools and businesses also began to emerge, and direct links to the media were being developed.

Tragically, just as the SOS Committee was making significant progress, it became a casualty of budget cuts as the Conservation Authority discontinued financial support and significantly reduced staffing in 1995. Provincial budget constraints were blamed; however, the Authority must also bear the responsibility because funding could have been reallocated from other areas within the LSEMS Implementation Program. The Authority made the decision assuming that the SOS Committee could continue to operate on its own. The SOS Committee, however, still comprised volunteers who had other commitments and, without the financial resources, had difficulty attracting other partners to deliver their programs. The loss of staffing resources probably played a more significant role than the loss of funding; the Committee absorbed many of the administrative duties and had less time to devote to implementation activities. Slowly, opportunities for the community to contribute toward rehabilitation efforts declined, and members became disillusioned and began leaving the Committee until it finally ceased meeting in the fall of 1995.

Efforts to revitalize the Committee were never realized because of the loss of provincial funding to LSEMS in 1996. It has been debated that the loss of the SOS Committee came at the time it was needed most, when community pressure could have possibly evoked a political response to address the loss of provincial funding. In any case, there is no question that loss of the SOS Committee was a significant blow to the LSEMS implementation efforts. During the beleaguered activities of Phase II, the communication and education component was significantly reduced to such a degree that there is a need to rebuild community support, and circumstances have been compared with the problems encountered during the initial stages of LSEMS Implementation Phase I.

The lessons learned from these experiences have not been lost on the Authority or the LSEMS partners. It is essential to involve the community in environmental programs like LSEMS, and substantial progress can be achieved through empowering the community to act. It is necessary though to realize that groups like the SOS Committee are made up of volunteers who, although they are willing to donate their time, have a limited supply. Therefore, just as agencies need support to maintain the sustainability of environmental programs, so do volunteer committees. The plans for Phase III of the LSESM Implementation Program include revival of the SOS Committee or establishment of a facsimile to assist with implementation by acting as a liaison to the community. Furthermore, the initiation of the State of the Watershed Study will also involve extensive public consultation and creation of a Community Advisory Committee to assist in the development of future resource management initiatives.

## CS5.7 Cost recovery and partnerships

In Chapter 5, the importance of developing partnerships to sustain long-term environmental monitoring programs is discussed. The need to continue monitoring programs within the Lake Simcoe watershed has been questioned more than once, especially when funding to complete capital projects all but disappeared. The importance of educating stakeholders regarding the value of collecting information over extended periods is essential to the continuation of any long-term monitoring program. Without long-term monitoring data collected under LSEMS, the relationship needed to establish goals and objectives and the plotting of trends to determine whether progress is being made toward achieving these goals would never have been developed. The complexity of natural systems also requires that long-term monitoring be conducted to discern the effects that other lake processes may have on phosphorus loading. Biotic interactions, invading species, and changes to the climatic conditions are some of the factors influencing phosphorus concentrations within Lake Simcoe.[198] The necessity of surveillance monitoring to detect new environmental problems is another rationale for long-term monitoring so resource managers can react promptly to address environmental concerns.

Throughout the LSEMS partnership, monitoring activities have primarily been the responsibility of the provincial agencies, and funding for these activities has always been available. Under the LSEMS Phase I and II agreements, the Ontario Ministry of Environment was designated to undertake water quality and quantity monitoring of Lake Simcoe and its tributaries, whereas the Ministry of Natural Resources was responsible for continued monitoring of the Lake Simcoe fish community. The Conservation Authority's main role as lead agency was to implement an interim phosphorus control management strategy and use the monitoring data to undertake predictive modeling exercises so all the partners could assist in the development of more effective remedial programs. The success of maintaining the provincial monitoring programs, despite significant time lags between the agreements and constraints imposed during a period of cost cutting and restructuring in 1997, in some way reflects the importance that the LSEMS partners placed on the monitoring information being collected.

During LSEMS Phase II, provincial restructuring and cost-cutting measures did impact the monitoring programs, resulting in the reduction of the sampling frequency of some of the tributary monitoring sites. It was recognized that the program would benefit from additional partners, and the regional and county municipalities were invited to join, as was another provincial agency, the Ministry of Municipal Affairs and Housing. The municipalities have long understood the importance of performance monitoring and have willingly provided additional revenues to augment existing monitoring efforts throughout the watershed. Opportunities to involve the private sector have not been attempted. Past relationships with the private sector have primarily involved obtaining support for the completion of environmental projects, not for what that sector considers "operational activities." However, in light of the successes being achieved elsewhere in the province (see Case Study 1), other LSEMS partners may be pursued should the need arise. Attempts to involve other external partners such as universities, colleges, and schools have also met with some success, and plans to enhance these partnerships will be initiated as part of Phase III. These partnerships were originally established to promote the use of data and create further awareness of the lake's pollution problems but have evolved to include in-kind support from students interested in assisting with data collection, input, and analysis.

The advent of GIS has also prompted the development of a new partnership within the LSEMS Monitoring Working Group, the creation of a user group to direct data collection efforts and ensure that data layers are updated to reflect the changing conditions within the watershed. The Regional Municipality of York has accepted the responsibility of warehousing the data and has already developed similar partnerships with the neighboring Regional Municipalities of Peel and Durham. The Province of Ontario and the Authority also have their own systems and will continue to develop data layers with the understanding that the Region of York may distribute the data to other partners in the user group. Issues regarding the ownership of data and distribution of information to nonpartners have yet to be resolved and will, we hope, be addressed in the near future. It is the opinion of the user group that the information collected is being done so with community tax dollars and should be made available free of charge to anyone who wants it.

## CS5.8 Conclusion

The experiences documented within this case study provide a convincing argument in support of the Closed-Loop Model described throughout this book. Ensuring the sustainability of any environmental program requires political support, sound scientific assessment techniques, an educated and involved community, and the development of partnerships and cost recovery programs. Each of the component parts is linked to one another, just as natural resources are linked within an ecosystem, and similarly, a balance between the components needs to be maintained. Repercussions will result should resource managers ignore one or more of these components, as has been demonstrated in this case study.

Over the last 10 years, the LSEMS partners have achieved success in maintaining the health of Lake Simcoe despite adversity and have learned some hard but extremely valuable lessons. There is still much more that needs to be done before the LSEMS goal of restoring a self-sustainable cold-water fishery in Lake Simcoe is realized. There is no doubt that the LSEMS Implementation Program, and the health of Lake Simcoe, could have benefited from an approach based on the Closed-Loop Model from the outset. It is vital that we learn from these mistakes and continue to improve our methods to achieve resource goals and objectives to sustain the health and quality of our natural resources for future generations.

# Case Study 6

# Applying the Closed-Loop Model to improve water quality in the Yuqiao Reservoir, Tianjin, China

## CS6.1 Summary

A pilot project to reduce water pollution using a public-participatory approach was developed for the Yuqiao Reservoir, the source of drinking water for the City of Tianjin in China. This project is one of nine model projects being executed by the Federation of Canadian Municipalities (FCM) and the Chinese State Council Office for Restructuring the Economic Systems (SCORES) under a bilateral agreement of the Government of Canada and the People's Republic of China. The project is supported by the Canadian International Development Agency (CIDA). The project team consisted of the FCM, International Centre for Municipal Development (ICMD), the Chinese SCORES, the Tianjin Environmental Protection Bureau (EPB), the Tianjin Academy of Environmental Science (TAES), the Ji County EPB, Ji County local government, and the Lake Simcoe Region Conservation Authority (LSRCA), a watershed management agency in Ontario, Canada. The project goal was "to improve the level of public involvement to reduce pollutants that contribute to poor water quality in the Yuqiao Reservoir." The role of the Canadian experts was to provide guidance and instruction to the project team to build capacity in public management for the various levels of the Chinese government. This was accomplished by assisting the project team to develop and implement public-participation components of pollution reduction plans in Ji County.

Three environmental rehabilitation projects were completed using the public-participatory process in the Chuanfangyu area of Ji County. The community was involved in all aspects of the projects and is at present participating in performance monitoring at each of the sites. The projects represent how all levels of the government and the community can work together to improve water quality in the Yuqiao Reservoir by adopting sound environmental practices. The benefits associated with involving the public were evident through the improved response of the community to undertaking and promoting the adoption of remedial measures within the pilot area. Government agencies have also recognized the value of the public-participatory approach, and the Tianjin EPB has continued to employ the process to assist in the completion of other environmental projects.

## CS6.2 Introduction

The FCM has been working in the international community since 1987. By establishing partnerships with municipalities in other countries, FCM is sharing Canadian expertise and technology to build management capacity into local governments worldwide. These initiatives encourage sustainable development by emphasizing processes that bring municipal governments and the community together in local decision making. In 1999, the Tianjin Model Project was initiated and is one of nine model projects being undertaken within the China Integrated Municipal Development Project (CIMDP). The project is funded by CIDA and is delivered by the FCM, ICMD, and the

Chinese SCORES. The LSRCA was contracted by FCM to provided instruction and assistance regarding public management methods used in Canada for pollution control.

The project involved working with all levels of government and the community to improve water quality in the Yuqiao Reservoir, which is experiencing severe pollution problems (i.e., cultural eutrophication and fecal bacterial loading) and is the main drinking water source for the City of Tianjin. Detailed investigations carried out by the City of Tianjin EPB in cooperation with the Ji County EPB identified that human activities within the watershed were responsible for the pollution problem. Specific sources of pollution included tributary rivers, agricultural activities, fish farming, human waste from villages, and soil erosion of the surrounding lands. The City acknowledged that past efforts to improve the water quality of the Reservoir had achieved only limited success and that a new and improved management approach was needed. To this end, a pilot project team was assembled to develop a plan to address the water pollution problems through improved public management and participation. The focus of this case study is the process used in developing the general public-participatory plan and the experiences of the project team to complete one of the three pilot projects.

## CS6.3 The public-participatory process: A means to building consensus

The key to improving the health of water resources is through a process of achieving social change.[64] As discussed in Chapter 4, the public-participatory process has long been recognized in Canada as the most effective approach to effect social change. The process works by involving the public in a strategic exercise designed to identify and solve problems through consensus building among stakeholders. Public involvement at the earliest stages of the project is the key to the success of the participatory process. Through involvement in the development of the projects, the public's role is transformed from that of observer into that of *stakeholder,* or a party with a personal interest in the success of the project. This step is essential so that the community accepts or buys into the plan, thus ensuring future cooperation during implementation.

In cooperation with the Tianjin pilot project team, FCM and LSRCA worked toward two objectives. The first was development of a general public-participatory plan for the delivery of a pollution control program for the Yuqiao Reservoir. The second related to building the management capacity of the various levels of Chinese government by teaching the stakeholder agencies how to use the public management approach for future application. Having the government agencies develop participatory plans for three pilot remedial projects in the Yuqiao Reservoir watershed completed this last task.

An eight-step process was used to develop the general public-participation plan for the Yuqiao Reservoir, as follows:

1. Identifying the problem
2. Developing a project goal
3. Defining the study area and pilot projects to be completed
4. Identifying objectives for involving the public
5. Outlining the benefits and obstacles to involving the public
6. Identifying the stakeholders
7. Outlining methods of public participation
8. Establishing an action plan

### CS6.3.1 *Identifying the problem*

The pilot project team identified that water quality deterioration in the Yuqiao Reservoir was originating from pollution from tributary rivers, agriculture and fish farming, habitation around the reservoir, and sedimentation and soil erosion. Because human land use activities and practices within the watershed area were largely to blame for the pollution inputs, the project team believed that efforts to effect social change would be required to address the problem at the source.

### CS6.3.2 *Developing the project goal and objective*

The Canadian experts worked through the planning exercise with the project team to introduce them to the facilitated management approach that they would need to use when working with the community. The project goal and objective developed by the team was to "improve the level of public involvement to reduce pollutants that contribute to poor water quality in the Yuqiao Reservoir." The project objective was to "build capacity for public management within government departments involved at various levels in Tianjin through developing a plan for public participation in activities to reduce pollution in a selected pilot area."

### CS6.3.3 *Defining the study area and pilot projects*

Chuanfangyu was chosen, and three remedial pilot projects were designed around the principle of public involvement with full cooperation from local stakeholders. The projects involved constructing control structures; reforesting mountainous areas to resist erosion; containing human waste and reducing other sources of contamination from villages; and changing land use and farming practices determined to be contributing pollution to the Yuqiao Reservoir. The projects were used for demonstration purposes to display practical solutions and promote wise stewardship of the watershed. They also acted as case studies to illustrate the effectiveness of public participation as a means of promoting uptake of best management practices.

### CS6.3.4 *Identifying objectives for involving the public*

The pilot project team saw the objectives for involving the public as follows:

- To improve public understanding of the nature of the water quality problems in the Yuqiao Reservoir and their societal implications
- To inform the public of the importance of improving water quality and obtain guidance on implementation of remedial works
- To enhance the cooperation and partnership within government departments and between authorities

### CS6.3.5 *Obstacles and benefits of using a public-participatory approach to control pollution*

Obstacles that could prevent the above objectives from being achieved were identified as part of the strategic process, as were plans to overcome them. Four major obstacles were identified. The first concern was the public's lack of formal education and awareness regarding the pollution problem of the Yuqiao Reservoir. To eliminate this obstacle, the first step of the public process was an educational initiative designed to boost awareness of the Yuqiao pollution problem.

The second anticipated obstacle was a lack of public trust in government. To overcome this obstacle, the project team concluded that the agencies and staff involved in the development and delivery of projects would have to work more closely with the community than in the past, providing technical and (where necessary) financial support. Consequently, agency officials were encouraged to be more approachable and responsive to the public's concerns. The development of closer ties between agency staff and the public ultimately built trust in the Tianjin region, which was vital to implementation of remedial projects.

The next obstacle concerned government agencies responsible for delivering the program. Given that the public-participatory process was new, there was some concern that there would be difficulties coordinating stakeholder involvement and that staff would not have the capacity to facilitate the public-participation process. Building capacity for public management was the responsibility of the Canadian experts and necessitated training within key agencies such that participants trained in the art of public management would ultimately be responsible for instructing other staff within the various levels of government so as to manage for succession.

An additional obstacle seen was the long-term nature of remedial projects and the risk that stakeholders would lose interest in the initiative if tangible benefits took some time to arise. This is a common problem when dealing with water quality rehabilitation programs because improvements may take years to materialize as pollutants that have accumulated within the natural system dissipate. Two solutions were proposed to overcome this obstacle. The first involved educating and informing the community about rehabilitation techniques, such that stakeholders would form reasonable expectations for project outcomes. The other solution involved performance monitoring of the project sites for water quality and other environmental and socioeconomic factors. In these instances, the community was asked to participate in the dissemination of results through discussions with neighboring communities.

Anticipated benefits of the public-participatory approach included the following:

- Improved environmental awareness and recognition of the problem affecting the Yuqiao Reservoir and of the environmental benefits associated with broad uptake of good stewardship practices would ultimately lead to enhanced quality of life within the community and opportunities for economic development.
- Remedial projects and proposed changes to farming practices would be made more practical (and hence be more widely implemented) by involving the public and farm community.
- Pilot projects would gain credibility from community involvement, and trust would be enhanced between the community and government agencies; participants would be more likely to promote the benefits of the projects in neighboring communities

## *CS6.3.6 Identifying the stakeholders*

Stakeholders were defined as individuals or agencies that would benefit from the pilot project. Two groups of stakeholders were identified including government agencies and members of the public, as described below.

The City of Tianjin EPB, a municipal government department, was identified as a stakeholder because of its involvement in environmental management and enforcement, pollution prevention and control (both in urban and rural areas), and monitoring.

The Ji County Local Government and the Chuanfangyu Village Government were primary partners in the project. The Ji County EPB (which has similar divisions as the Tianjin EPB at a local level) was an obvious stakeholder and took on a role of stakeholder coordination in the pilot project. The Water Conservation Authority, a municipal government department in Tianjin, was identified as a stakeholder because of its involvement in flood control, drinking water supply (quantity and quality), water conservation, data collection, and monitoring. The local Agricultural Authority (a municipal government department that is involved in land use planning related to agriculture), Forest Authority (a municipal government department involved in forest management), Mining and Geological Authority (a municipal government department involved in erosion control along steep slopes and planning mineral and aggregate extraction), and Education and Publicity Authority (a municipal government department involved in the education of children) were similarly identified as stakeholders.

Public interest groups invited to participate in the pilot project included the Youth League (a youth organization involved in government-requested activities and volunteerism), the local Women's Federation Organization (an organization representing women's affairs and rights and represented at all levels of government), and the Village Committee (a grassroots government administrative council organization coordinating farming and other village activities, and individual farmers).

### CS6.3.7 Outlining methods of public participation

Once stakeholders were identified and invited to participate, the next step in the development of the participation plan was to establish the methods to be employed to educate the public and facilitate involvement.

Meetings and workshops were held in the community to educate and train the public regarding the benefits of stewardship practices and to solicit their comments and support. The local media (television, radio, newspapers) were used to promote the workshops and other events and also as a secondary means of delivering educational messages. Specific educational newsletters and fact sheets were distributed to local households and posted on community bulletin boards.

Surveys garnered feedback from stakeholders and were used as a performance measure for the public involvement initiative. Financial incentives were provided to local farmers to boost uptake of remedial projects. Technical support was provided to producers willing to adopt new farming practices and initiate environmental projects. A participant recognition program was developed, including an awards event to be scheduled periodically to honor good stewardship practices and the people involved in employing them. Finally, a scientific assessment program was developed to monitor environmental benefits associated with the projects implemented through the public process.

### CS6.3.8 Producing a public-participation action plan

The final step in developing the public-participation pilot was to produce a feasible, goal-oriented action plan. The action plan focused on three main aspects of the pilot project:

1. Public management, that is, how the project team would manage and execute its efforts
2. Project development, that is, how the project team would select and implement individual remedial works such as modifying cropping practices, controlling erosion, and managing urban wastewaters
3. Maximization of opportunities for stakeholder participation in each facet of the pilot project public participation: the opportunities to involve stakeholders in improving water quality

By 2000, it was evident from a review of the pilot projects by the Canadian experts that the pilot project team had made significant progress in developing public management and project development skills necessary to implement the participation plans. After the public-participation process, stakeholders from all levels of government and differing departments, from the City to the Village levels, had been organized to work together. The eight-step participatory process had also been followed for each of the projects. Actions taken to execute each of the plans to complete the pilot projects had been documented, and most timelines established had been met. The most significant achievement observed by the Canadians was that the public had indeed been provided the opportunity to participate in the decision-making process as well as the implementation of individual remedial projects. Because of their mutual involvement in the project, the community and the government agencies were brought together and developed a cooperative relationship based on trust.

## *CS6.4 Achieving success: Putting the plans into action*

One project in particular provided an excellent example of the achievements of the pilot project team in applying the public-participatory process. The project involved working with the farming community to promote the planting of a cover crop between rows of grapes in the village of Yaobaizhunag. The problem identified by the Tianjin project team was that "the poor soil structure and planting of grapes parallel to slope was resulting in increased runoff and pollutant loading entering the Yuqiao Reservoir." The project team proposed using a cover crop between the grape rows to protect the soil surface, restore soil structure, and ultimately reduce runoff from the field. The reduced runoff would result in improved water quality by decreasing soil particles and fertilizers entering the Yuqiao Reservoir.

A switch from growing corn and wheat to cultivating grapes was gaining popularity in the area because of the increasing demand for fruit and wine production. An added benefit was that the space between the rows of grapes could still be used to cultivate some crops, reducing the loss of income associated with taking land out of production. As a result, farmers would increase their revenue in the long term while sustaining it in the short term. This trend to change crop production and grow grapes was increasing within the watershed of the Yuqiao Reservoir and posed a threat to the pilot project objective of improving water quality in the Yuqiao Reservoir. As a result, the project team identified that measures must be taken to reduce pollution inputs associated with the shift in agricultural production. The resultant project goal was "to work with the farming community to develop and implement changes to existing cropping practices to reduce water quality degradation."

With the problem identified, an interim project goal established, and a study area selected (Yaobaizhuang Village), the next step in the process was developing a methodology for public consultation. A local village leader was approached to assist members of the project team in this step. The leader was asked to identify likely farmer candidates and eventually to introduce members of the team to the potential farmers who might be willing to participate. Individual meetings were held with the candidate families to discuss the project and determine their willingness to get involved. Meetings were also used to inform the farmers of the water quality problem, obtain feedback on farming practices, and solicit landowner participation. Printed information describing the problems and the potential solutions (fact sheets) were produced and provided on-farm. Survey sheets documented landowner feedback.

Obstacles identified during the producer meetings included concerns over potential lost income resulting from the implementation of soil conservation measures (both remedial solutions identified above would result in a period of lost revenue because cropland would—at least temporarily—be taken out of production). This issue had been anticipated as the most significant obstacle that would be encountered and was addressed through financial incentives equal to the revenue that would be lost while the fields were out of production. Benefits associated with the proposed projects were also identified, including improved soil structure, reduced runoff and water pollution, as well as the fact that the economic benefits associated with grape production would far exceed the revenues from continued corn or wheat production, based on existing markets. As a result, the farm community was much more inclined to adopt the change to soil conservation–oriented grape production than to return to producing corn and wheat.

The Yaobaizhuang pilot project stakeholders included the following:

- Ji County Local Government
- Tianjin Academy of Environmental Science
- Tianjin Environmental Monitoring Centre
- Ji County Environmental Protection Bureau
- Town of Chuanfangyu
- Agriculture Station of Chuanfangyu
- Village Leader of Yaobaizhuang Village
- Farmers of Yaobaizhuang Village

As the local environmental organization, the Ji County EPB was designated as the lead agency responsible for managing the overall project and coordinating the work of the other government agencies. The Tianjin Academy of Environmental Science, in cooperation with the local agencies and politicians, developed the educational material and assisted in their distribution and in initial meetings and presentations to the community. After these educational steps were taken, meetings were held to discuss the proposed project with a number of potential candidates. Feedback from the farm community was collected at this stage, and input was provided on how the proposed project could be improved. Efforts were made to incorporate suggestions made by the public, and these resulted in one family volunteering to establish cover crops on one cultivated grape field. The family was compensated for the loss in revenue that the land taken out of production would have yielded. With all the obstacles addressed, the project was initiated, and the family, in cooperation with representatives from the Agriculture Station of Chuanfangyu, undertook the task of seeding the selected cover crop (red clover) and monitoring its progress.

As might be expected, the project elicited a great deal of interest from other farmers within the village and the Chuanfangyu area. Because of the efforts of the project team in following the participatory process, the farmer had made the transition from observer to stakeholder. In this role, the farmer had also taken on the added responsibility of educator, to promote the change in farming practice to other farmers within the area. Realizing the added benefit of promoting the project further, staff from the Agriculture Station of Chuanfangyu and the Ji County EPB would routinely bring farmers from other villages to visit the site and to meet with the farmer to share in his experience.

Response to the project was decidedly mixed. The main concern expressed by the farmers visiting the site was the need to observe the benefits of the project before widespread application of the practice. This obstacle had been identified in advance by the project team, and an appropriate performance-monitoring program to document the benefits had been developed and was being implemented by the Tianjin Environmental Monitoring Centre. These results would take time to obtain (2 to 3 years), and a commitment was made to the farming community to keep it informed of the project progress. Regardless of their response, visiting farmers witnessed firsthand one of the most important benefits of the participatory process—that as a result of the government and the community working together on the project, a solid relationship built on trust had been established, especially with the farmer.

## *CS6.5 Lessons learned and the continued application of the public-participation process in China*

The experiences of the Project Team showed that a participatory approach to water resource management can be successfully used in less developed and developed countries, as is done commonly in North America and elsewhere. The Chinese government agencies demonstrated that they could build capacity in public management and project development skills necessary to facilitate the participatory approach. In addition, the Tianjin pilot project demonstrated that stakeholder groups and various levels of government can work together to achieve environmental and economic benefits.

Proper planning and agency support can surmount obstacles associated with involving the public, as observed from the three pilot projects, in which the Chinese public eagerly became involved in the decision-making process.

By educating the public regarding the pollution problems and providing solutions and opportunities for public involvement, progress to improve water quality and the general quality of life within the area of the Yuqiao Reservoir has been achieved, and similar approaches could be used elsewhere. Listening and incorporating local concerns and issues into projects will improve public acceptance of proposed rehabilitation measures and increase implementation within the community. By working together to create relationships based on mutual trust, closer ties between the community and government agencies can also be established. To develop this rapport within the community, government agencies must do the following:

- Involve all the stakeholders including the community at the very beginning of the project
- Ensure that projects to be completed are mutually beneficial to all stakeholders
- Encourage the community stakeholders to take on roles of educators and talk about what they had learned through their own experiences
- Provide the appropriate technical and financial support

Good public management is also critical to the successful implementation of the participatory process. As discussed in Chapter 5, partnerships work because they bring together organizations and people with a wide range of experience and expertise. The creation of a project team comprising the appropriate stakeholders is essential to the success of each project, as is the development of an action plan and designation of a lead agency responsible for implementing the plan. Ensuring that the staff involved in the project have the proper training and project development skills is another contributing factor to the success (or failure) of a project.

Some of the most satisfying outcomes of the project for the Canadian experts were learning that the Project Team would continue to use the participatory process to address water quality concerns within the Yuqiao Reservoir after the termination of overseas involvement and, furthermore, that the Tianjin EPB and Academy of Environmental Science had begun to adopt the public-participatory management approach to complete other environmental projects. In Canada, an environmental assessment process has been legislated at a federal, provincial, and municipal level to review projects that have the potential to negatively impact the environment. This process is loosely based on a public-participatory approach and has ensured that environmental impacts associated with the development of infrastructure (highways, dams, development of waste management sites, etc.) are mitigated as much as possible. Using a very similar approach, the Tianjin Academy of Environmental Science applied the public-participatory process to develop solutions and alleviate the community's concerns regarding the Blue Mountain Energy Project. By involving the public in this type of project, the Tianjin EPB may well have embarked on a new approach toward environmental management. At the very least, the continued use of the participatory approach demonstrates that the Tianjin Pilot Project was successful in influencing public management methods in local decision making.

FCM and the project participants gratefully acknowledge the support provided by CIDA for its international programs and publications.

# Glossary

**accuracy** a measure of how closely results match true or expected values

**adaptive management** a cyclic management program in which strategies are modified on a periodic basis according to results of performance monitoring

**ANOVA** acronym for *analysis of variance*, a standard statistical tool for determining significance of results during hypothesis testing

**antagonistic mechanism** relating to the combined environmental effect of two contaminants acting in concert being of lesser magnitude compared with the effects of the same contaminants acting independently (destructive interference)

**antecedent moisture** the degree of wetness of the soil at the beginning of a runoff period, frequently expressed as an index determined by the summation of weighted daily rainfall for a period preceding the runoff event in question

**antecedent precipitation index (API)** index of moisture stored within a drainage basin before a storm

**anthropogenic** human induced

**aquifer** underground waterbearing formation; saturated, permeable geologic unit that can transmit groundwater according to prevailing hydraulic gradients

**ARC/INFO** geographic information system software by ESRI, Inc.

**area of concern** one of a number of locations in both Canadian and U.S. jurisdictions within the Great Lakes Basin deemed high priority for remediation of water quality and habitat impairments by the International Joint Commission

**assimilation** maintenance of aquatic ecosystem function despite additions of contaminants (nutrients, pesticides, metals, etc.) through processing and sequestration into biomass and other compartments of the system

**assimilative capacity** relative measure of the capacity of aquatic systems to process contaminants (often phosphorus and other human wastes) and to sequester them in ecologically benign forms and environmental compartments

**base flow** estimate of the minimum discharge of a stream during the driest time of the year, a time at which most of the flow in the channel arises from groundwater discharge into the channel rather than from runoff or direct inputs from precipitation

**benthos** bottom-dwelling (benthic) aquatic invertebrates; this group of taxa includes arthropods (insects, crustaceans, etc.), annelids (worms, leeches, etc.), and others

**best management practice (BMP)** a technology that promotes watershed health, usually by improving some ecological feature, and requires minimal social change or financial investment

**bioaccumulation** process in which uptake of contaminants into biomass results in higher contaminant concentrations in biota than in water or other compartments of their habitat

**biocriteria standards** parameters, usually for water quality, based on biotic community composition

**biodiversity** used interchangeably with *taxa richness,* the total assemblage of taxa groups present at a given location and time

**biological assessment, bioassessment** approach using biological responses to characterize the degree to which human activity has shaped the ecology of an area (commonly streams or other bodies of water)

**biological integrity** at any given spatial scale, a measure of the ability of a habitat to support an assemblage of taxa that is comparable to that of similar, undisturbed habitats

**biomagnification** process by which the concentration of a contaminant reaches sequentially higher concentrations in biota of higher trophic levels

**biotic index** common measure of biological community health; calculation usually involves richness and abundance measures combined with sensitivity values (that reflect empirically derived responses to the disturbance regime in question)

**Boolean operators** symbols used to specify logical operations in an expression, for example, *AND, OR,* ≥

**bottom-up process** governance model in which primary decision-making influences are channeled to management bodies from the community (contrast with *top-down process*)

**carrying capacity** theoretical concept describing the maximum amount of human presence or activity that can take place in a watershed (or other area) while ecological function (watershed health) is maintained

**channel routing** model simulation of stream flow, usually used to predict water levels (e.g., in flood forecasting), erosive energy, or travel time between points in a catchment along a flow pathway

**class environmental assessment** standardized study format used to predict environmental impacts associated with a specific land use change

**cold-water stream** groundwater-rich stream, usually with high base flow yield per unit catchment area, that provides tremendous thermal buffering (hence relatively little diurnal and seasonal variation in temperature); these systems have characteristic taxa assemblages and water chemistry and, in North America, support important trout and salmon fisheries

**comparability** measure of the variance between data collected at different locations or times; may also relate to the similarity of two or more methods for data collection, analysis, or reporting

**completeness** measure of the difference in the amount of valid, or usable, data collected during a study compared with the quantity called for in the experimental or survey design

**compliance monitoring** type of surveillance monitoring that compares ambient conditions against established regulatory or government standards or criteria

**covariance** correlation between independent variables

**dependent variable** response or outcome variable (e.g., abundance, richness, or growth rate)

**digital elevation model** computer-generated database of ground elevations and topographical contours, referenced to a common datum

**digital terrain model** three-dimensional computer-generated representation of a ground surface using databases of digital elevations and other terrain features and properties

**direct approach** exact measure of some environmental (or other) attribute such that the result is not subject to sampling error

**distributed watershed model** hydrologic model that applies a grid pattern across the watershed landscape as a means of discretizing the area; estimates of runoff (or other hydrologic parameters) over larger areas are made by integrating calculated values across the appropriate cells in the grid;. models can usually aggregate grid cells into *homogeneous hydrological units* (HHUs) to yield a higher computational efficiency; the choice of running the model in either HHUs or on a cell-by-cell mode is application dependent and is generally a function of basin size as well as of the expected temporal and spatial variations

**diversity index (community structure index)** univariate measure of community health based on the ratios of abundances of individual taxa compared with total counts for the site. A combination richness and abundance measure

**ecological functions** natural processes governing the sustainable provision of life requisites; include, but are not limited to, dispersal, decomposition, nutrient cycling, and maintenance of genetic variation and selection

**ecological integrity** measure of how closely ecosystem function and biodiversity meet physiographic potential; this term is generally used on a watershed scale but may be applied to subwatershed or site scales

**ecosystem** derived from the Greek *oikos* (meaning "household") and *system* (a complex of interdependent elements), it is the lowest ecological unit that provides all life requisites; it is not a closed geographic unit, but an open unit with inputs and outputs[5]

**end-of-pipe solution** treatment process used on effluent in a location remote from the source of contamination.; typical examples include sewage treatment plants and constructed wetlands that often treat effluent collected from large areas and deliver a treated product to a specific location within a surface water body

**eutrophication (cultural)** increase in productivity of a water body as a result of anthropogenic inputs of phosphorus or other nutrients

**evapotranspiration** process by which plants give off water vapor to the air, a combination of evaporation and transpiration

**expert system** assessment methodology in which the differentiation between impaired and unimpaired condition is largely based on expert interpretation of data

**flood forecasting** early warning system implemented in populated watersheds to reduce or prevent human and financial costs associated with flood events; usually relies on weather forecasting, real-time monitoring of precipitation events, and stream flow, along with model information, to predict water levels in flood-prone areas

**flood lines** map elevation contours representing the water level associated with floods of a given magnitude or resulting from a precipitation event of specific magnitude or frequency of occurrence

**floodplain** land that is adjacent to a stream channel and is subject to flooding when the stream overtops its banks; in hydrologic terms, it is the area subject to inundation by floods of a particular frequency (10-year, 20-year floodplains, etc.)

**flood-vulnerable area (FVA)** area susceptible to flooding based on a rainfall event of given severity or frequency of occurrence

**flow node** in relation to hydrology modeling, a point of interest along a watercourse system where stream flow properties are simulated; typical examples of flow nodes are stream confluences, bridges, and culverts

**geographic information system (GIS)** automated (usually computer-based) system capable of compiling, storing, manipulating, and displaying spatial (geographic) and temporal information;[123] the primary role of a GIS is to link spatial mapping data with database records of attributes that correspond to segments of map data

**grab sample** simple water chemistry sampling procedure whereby practitioners manually fill sample containers for laboratory submission

**groundwater** water partitioned into the subterranean zone; one considerable storage compartment within the hydrologic cycle

**hydrogeology** the study of the nature and distribution of groundwater within geologic systems

**hydrologic cycle** the endless flux of water (in its various physical states, or phases) among atmospheric, surficial (oceans, lakes, rivers, etc.), and subterranean zones

**hydrologic/hydrogeologic model** (generally computer-based) means of estimating groundwater or stream flows, water surface elevations, and other water system parameters at selected locations in the watershed under examination

**hydrology** the study of the occurrence, distribution, movement, and properties of water in the environment

**impaired** status of aquatic health in which human activities (land use, pollution, etc.) are the main factors shaping habitat, water quality, and the aquatic community (contrast wth *unimpaired*)

**independent variable** variable that is controlled or considered fixed and that may affect the value of a response variable (common examples include time or location)

**indicator (environmental)** measurable feature that provides reliable evidence of watershed health and can therefore be used as a tool for surveillance or performance evaluation

**induced recharge** action of facilitating recharge to groundwater; this is often done as a best management practice in conjunction with land development works (e.g., infiltration trenches, perforated sewer pipes, reduced lot grading, etc.) to reduce the volume of storm water runoff generated and requiring treatment

**infiltration** the movement of water through permeable soil layers down to the water table

**Integrated Stormwater and Watershed Management System (ISWMS)** proprietary hydrologic software package by Greenland International Consulting Inc. that is used for managing water resources in urban or rural watersheds

**interception** typically used in the context of a short circuit in the hydrologic cycle; for example, installation of perforated drainage tiles in an agricultural field represents an interception because the localized patterns of infiltration are short-circuited as a portion of the water that penetrates the ground's surface is shunted directly into a surface water receiver

**isomer** molecule with the same chemical formula but different three-dimensional structure as another (a structural homologue)

**lentic** relating to a standing body of water

**living document** document (often used in planning exercises) containing a schedule for periodic review and amendment (i.e., of targets or strategies) developed through an interactive public process

**loading, pollutant** measure of the amount of a given contaminant carried downstream past a given point along a river or stream; typical units of measure include tons per year, milligrams per second, and so on

**lotic** relating to a flowing body of water

**lumped watershed model** hydrologic model that incorporates a homogeneous application of modeling algorithms across a user-defined basin

**model** (1) scaled reproduction or representation of an entity, treatment process, or environmental phenomenon; (2) mathematical estimate of an environmental parameter

**modeling** simulation of some physical or abstract phenomenon or system through manipulation of a simpler, analogous system that is assumed to obey the same physical laws or abstract rules of logic

**MODFLOW™** acronym given to the U.S. Geological Service's Modular Three-Dimensional Groundwater Flow Model; has become an industry-standard groundwater flow model, with its ability to simulate a wide variety of systems, its extensive publicly available documentation, and rigorous peer review; can be used to simulate groundwater systems for water supply, contaminant remediation, and mine dewatering and is used by regulatory agencies, universities, consultants, and industry

**multiple regression** statistical approach used to sequentially improve predictions of the value of a response variable by inclusion of additional independent variables into a regression model

**objectives** general claims about the relation of a given watershed management activity to watershed health, for example, "the objective of our flood forecasting program is to prevent property damage and loss of life"

**organic** as in *organic contaminant*, a carbon-based chemical; an example of an organic contaminant is DDT (dichlorodiphenyl-trichloroethane)

**organochlorines** group of synthetic chlorine-containing organic compounds that are problematic in aquatic systems because of their persistence or longevity in the environment (the chlorine decreases the rate of decomposition in the environment) and also their toxicity. They tend to be insoluble in water (hydrophobic) and readily stored in fat (lipophilic)

**partnering model** cost recovery model in which cash and in-kind support are exchanged in a mutually beneficial way between the lead watershed group and a variety of stakeholders. Typically, cash and in-kind resources flow toward the lead watershed agency to sustain the program, whereas information, influence, and credibility are bestowed upon involved stakeholders

**performance monitoring and evaluation** assessment of the progress made toward objectives (stated in the watershed plan) through implementation of management policies and strategies

**periphyton** bottom-dwelling algae that are attached to the substrate

**pollutant** anthropogenic contaminant (often of waters)

**pollutant trading** pollutant management framework in which contributors of a given contaminant are given quotas for discharge of pollutants into the environment in exchange for remediation of another source of the same pollutant in a remote location; generally proceeds on a net-gain basis, whereby one contributor may be granted a permit for $x$ mg/day in exchange for financing remediation of $4x$ mg/day at another location in the watershed

**precision** degree of agreement among repeated measurements of the same characteristic (i.e., variance)

**quality assurance** overall strategy for watershed health monitoring that ensures that monitoring meets managerial requirements; includes procedures for data collection, quality control, documentation, evaluation, and reporting

**quality assurance plan** written document outlining the procedures that a monitoring project will use to ensure the data it collects and analyzes meet project requirements

**quality control** set of procedures aimed at minimizing the occurrence of errors in data sets; generally composed of guidelines for field, laboratory, and office procedures to minimize occurrence of errors; a subcomponent of quality assurance

**rapid bioassessment protocol** family of biological assessment techniques that typically use a reference condition approach and employ sampling and analytical techniques that are conducive to rapid screening of sites for indications of impairment (coarse taxonomic level, subsampling, or fixed counts during sample processing; comparison of compositional indices against thresholds to highlight sites that are atypical in relation to the reference condition; etc.)

**reference condition approach** assessment methodology (commonly employed in bioassessment studies) in which impairment is measured as deviation in community composition of a test site in relation to a group of reference sites known to be minimally impacted by human activities or representing a desirable biotic condition; typically a multivariate approach with replication at the site level, rather than at the within-site level[23]

**reference site** one of a set of locations used to characterize the range of biological, chemical, and physical attributes present at minimally impaired sites; this range of conditions is used as a benchmark for assessing test sites in a reference condition approach[23]

**representativeness** extent to which measurements of some attribute actually represent the true environmental condition or population at the time of sampling

**SMART target** specific, measurable, attainable, relevant, and time-bound endpoint associated with a watershed management activity

**standards** policy guidelines including maximum allowable concentrations (e.g., of pollutants) or other thresholds generally related to the suitability of water for drinking, industrial processes, habitat, or other risk management considerations

**steady-state backwater model** computer program that estimates the longitudinal slope of the water surface in a stream or open conduit; these models are typically used to estimate upstream areas of impact associated with artificial constrictions or blockages (e.g., dams and weirs) in the channel

**STORET** water quality data management and storage system managed by the U.S. Environmental Protection Agency

**storm water** runoff water generated from precipitation onto impervious surfaces

**stormwater management facilities** treatment works used to remove suspended solids, oil and grit, nutrients, and other contaminants from stormwater

**stream health** statement of how closely the living stream community (i.e., benthic invertebrates, fish, etc.), water quality, and stream morphology at a test site match these same variables as measured at reference sites in a similar geographic and physiographic region

**surrogate (parameter)** easily measured parameter that is presumed or shown to be strongly correlated with some fundamental but difficult to measure environmental (or other) indicator; for example, consider a situation in which one is interested in assessing the quality of life in a particular city; the problem here is that quality of life is a subjective, abstract parameter that is difficult to measure, and for this reason, one may choose to measure one or more surrogates that are believed to reflect the quality of life; these may include the number of families that own cars or the proportion of people that are employed (both correlated with affluence), the proportion of the landscape that is forested (linked to aesthetics and recreational opportunity), or the murder rate (linked to personal safety and community morale)

**surveillance** or **reconnaissance** type of monitoring aimed at assessing watershed heath and how it varies across time; typically, surveillance monitoring activities focus on assessing environmental health, but the term is equally applied to measurements of watershed attributes on the human side of the watershed health equation (see Figure 1.1)

**sustainability** state of dynamic equilibrium in which human needs and ecological function are in balance and, hence, there is no net long-term reduction in watershed health; a sustainable management practice is one that promotes this balance

**synergistic** relating to the combined environmental effect of two contaminants acting in concert exceeding the effects of the same contaminants acting independently (constructive interference)

**taxa richness** number of taxonomic groups (often species) present; this term is used at different spatial and temporal scales; that is, one may refer to taxa richness from a site or from a subwatershed, or one may compare richness from one year to another

**top-down process** governance model with primary decision-making influences channeled from elected or appointed management bodies to the community (contrast with *bottom-up process*)

**Total Maximum Daily Load (TMDL)** a regulatory compliance threshold set according to the maximum daily pollutant contribution allowed from a given catchment area

**unimpaired** status of aquatic health in which natural processes are the major influence on habitat, water quality, and the aquatic community (contrast with *impaired*)

**variance** statistical measure of central tendency (the expected squared deviation of a probability distribution from its mean); an important tool for estimating the precision of repeated measures and sampling and experimental error

**visual OTTHYMO** hydrologic and stormwater management program developed by Greenland International Consulting Inc.; version 2.0 was released in early 2001

**water budget** or **water balance** estimation of the partitioning of water into the various compartments of the environment (i.e., atmospheric and terrestrial zones) and the rate of flux between them; typically, computerized hydrologic models are employed to estimate the water budget or balance for an area considering a number of parameters, including precipitation (rain, snow, etc.), runoff, snowmelt, infiltration and groundwater recharge, and evapotranspiration; traditionally, watershed managers have undertaken water budget modeling primarily in developed watersheds as a means of developing water allocation frameworks to avoid conflict over competing uses

**watershed** surficial drainage basin

**watershed health monitoring** methodology for assessing the balance between human uses and ecological function of a watershed

# References


1. Suzuki, D. 1994. *Time to Change Essays.* Stoddart Publishing Co. Ltd, Toronto, Ontario, Canada.
2. Workshop Report: International Workshop on River Basin Management, the Hague, the Netherlands, 27–29 October 1999. *Recommendations and Guidelines on Sustainable River Basin Management.* Available at: http://ct.tudelft.nl/rba/recommendations.html.
3. U.S. Environmental Protection Agency, Office of Wetlands, Oceans, and Watersheds. *An Introduction to Water Quality Monitoring.* Available at: http://www.epa.gov/owow/monitoring/monintro.html.
4. United Nations, Population Division, Department of Economic and Social Affairs. *1998 Revision of the World Population Estimates and Projections.* Available at: http://popin.org/pop1998/.
5. Odum, E.P. 1998. Foreword. In *Watershed Management Practice, Policies, and Coordination,* R.J. Reimold, Ed., McGraw-Hill, New York, pp. xiii–xiv.
6. North Carolina State University. *Michigan—Sycamore Creek Watershed—1996, 319 Report.* Available at: http://h2osparc.wq.ncsu.edu/96rept319/MI-96.html.
7. Land Conservancy of San Luis Obispo County. *San Luis Obispo Creek Forum.* Available at: http://www.callamer.com/landcon/taskforce/links.htm.
8. Racoon River Watershed Project. *Racoon River Watershed Project.* Available at: http://www.rrwp.org/.
9. Clean Water Action Plan. *The Upper and Lower Bad River Watersheds—Addressing Water Quality through Land Management.* Available at: http://cleanwater.gov/success/bad.html.
10. Department of Natural Resources. *King County Washington—Water and Land Resources.* Available at: http://dnr.metrokc.gov/wlr/wlrtopic.htm.
11. Golden, B.F. 1998. Issues in developing and implementing a successful multiparty watershed management strategy. In *Watershed Management Practice, Policies, and Coordination,* R.J. Reimold, Ed., McGraw-Hill, New York, pp. 353–368.
12. Browner, C.M. 1996. A watershed approach framework. In *Watershed Management Practice, Policies, and Coordination,* R.J. Reimold, Ed., McGraw-Hill, New York, pp. 369–384.

13. U.S. Environmental Protection Agency, Office of Wetlands, Oceans and Watersheds. *An Introduction to Water Quality Monitoring*. Available at: http://www.epa.gov/owow/monitoring/monintro.html.
14. U.S. Environmental Protection Agency. *Top 10 Watershed Lessons Learned.* Available at: http://www.epa.gov/owow/watershed/lessons/index.html.
15. U.S. Environmental Protection Agency. *Draft Framework for Watershed-Based Trading.* Available at: http://www.epa.gov/owow/watershed/framwork.html.
16. United States Department of Labor. *Why Is Consensus Difficult to Achieve?* Available at: http://www.dol.gov/dol/asp/public/futurework/conference/nonunions/concensus.htm.
17. International Alert. *Capacity Building Workshops.* Available at: http://www.international-alert.org/pdf/respk_section3.pdf.
18. Beck, C.H. 1973. J.W. Goethe, Werke, Hamburger Ausgabe, Munich. As translated in *Theater, Theory, Speculation: Walter Benjamin and Scenes of Modernity, Baltimore and London,* R. Nägele, 1991, Johns Hopkins University Press, Baltimore, MD.
19. U.S. Geological Survey. *The Strategy for Improving Water Quality Monitoring in the United States—Summary.* Available at: http://water.usgs.gov/wicp/summary.html.
20. Jones, C. *Stream Restoration Monitoring Framework.* Available at: http://www.nvca.on.ca/monitoring/stream/index.htm.
21. Yoder, C.O. *Important Concepts and Elements of an Adequate State Watershed Monitoring and Assessment Program.* Available at: http://nwqmc.site.net/98proceedings/papers/59-yode.htm.
22. Barbour, M.T., J. Gerritsen, B.D. Snyder et al. 1999. *Rapid Bioassessment Protocols for Use in Streams and Wadeable Rivers: Periphyton, Benthic Macroinvertebrates and Fish,* 2nd ed., U.S. Environmental Protection Agency, Office of Water, Washington, D.C.
23. Norris, R. 1996. Predicting water quality using reference conditions and associated communities. In *Proceedings of the 44th Annual Meeting of the North American Benthological Society, 9th annual Technical Information Workshop,* North American Benthological Society, Kalispell, MT, pp. 32–52.
24. Cairns, J., Jr. and J.R. Pratt. 1993. A history of biological monitoring using benthic macroinvertebrates. In *Freshwater Biomonitoring and Benthic Macroinvertebrates,* D.M. Rosenberg and V.H. Resh, Eds., Chapman & Hall, New York, pp. 10–27.
25. American Public Health Association. 1985. *Standard Methods for the Examination of Water and Wastewater,* 16th ed., American Public Health Association, Washington, D.C.
26. Cairns, J., Jr. and K.L. Dickson, Eds. 1973. *Biological Methods for the Assessment of Water Quality,* Special Technical Publication 528, American Society for Testing and Materials, Philadelphia.
27. Griffiths, R.W. 1993. *BioMAP: Concepts, Protocols and Procedures for the Southwestern Region of Ontario.* Ontario Ministry of Environment and Energy, London, Ontario, Canada.
28. Hellawell, J. 1977. Biological surveillance and water quality monitoring. In *Biological Monitoring of Inland Fisheries,* J.S. Alabaster, Ed., Applied Science Pubs., U.K., pp. 69–88.

29. Hellawell, J.M. 1978. *Biological Surveillance of Rivers: A Biological Monitoring Handbook.* Water Research Centre, Medmenham, U.K.
30. Hynes, H.B.N. 1960. *The Biology of Polluted Waters.* Liverpool University Press, Liverpool, U.K.
31. James, A. and L. Evison, eds. 1979. *Biological Indicators of Water Quality.* Wiley, Chichester, U.K.
32. Lenat, D.R., L.A. Smock, and D.L. Penrose. 1980. Use of benthic macroinvertebrates as indicators of environmental quality. In *Biological Monitoring for Environmental Effects,* D.L. Worf, Ed., D.C. Health, Lexington, MA, pp. 97–112.
33. Rosenberg, D.M., H.V. Danks, and D.M. Lehmkuhl. 1986. Importance of insects in environmental impact assessments. *Environmental Management* 10:773–783.
34. Rosenberg, D.M. and V.H. Resh. 1993. Introduction to freshwater biomonitoring and benthic macroinvertebrates. In *Freshwater Biomonitoring and Benthic Macroinvertebrates,* D.M. Rosenberg and V.H. Resh, Eds., Chapman & Hall, New York, pp. 1–9.
35. Wiederholm, T. 1980. Use of benthos in lake monitoring. *Journal of the Water Pollution Control Federation* 52:537–547.
36. Worf, D.L., Ed. 1980. *Biological Monitoring for Environmental Effects.* D.C. Health, Lexington, MA.
37. Griffiths, R.W. 1998. *BioMAP: A How to Manual.* Ministry of Municipal Affairs and Housing, Toronto, Ontario, Canada.
38. Ohio Environmental Protection Agency. 1987. *Biological Criteria for the Protection of Aquatic Life: Volumes I–III.* Ohio Environmental Protection Agency, Columbus.
39. Ontario Ministry of the Environmnent, Environmental Bill of Rights Office. *EBR Related Acts and Regulations.* Available at: http://www.ene.g.ov.on.ca/envision/env_reg/ebr/acts%20and%20regs/.
40. National Council for Science and the Environment. *Summaries of Environmental Laws Administered by the EPA—Clean Water Act II.* Available at:http://www.cnie.,org/NLE/CRSreports/BriefingBooks/Laws/.
41. Norris, R.H. and A. Georges. 1986. Design and analysis for assessment of water quality. In *Limnology in Australia,* P. De Deckker and W.D. Williams, Eds., Commonwealth Scientific and Industrial Research Organization, Melbourne and Junk Publishing, Dordrecht, the Netherlands, pp. 555–572.
42. Johnson, R.K, T. Wiederholm, and D.M. Rosenberg. 1993. Freshwater biomonitoring using individual organisms, populations and species assemblages of benthic macroinvertebrates. In *Freshwater Biomonitoring and Benthic Macroinvertebrates,* D.M. Rosenberg and V.H. Resh, Eds., Chapman & Hall, New York, pp. 40–158.
43. Hawkes, H.A. 1979. Invertebrates as indicators of river water quality. In *Biological Indicators of Water Quality,* A. James and L. Evison, Eds., Wiley, Chichester, U.K., chap. 2.
44. Resh, V.H. and E.P. McElravy. 1993. Contemporary quantitative approaches to biomonitoring. In *Freshwater Biomonitoring and Benthic Macroinvertebrates,* D.M. Rosenberg and V.H. Resh, Eds., Chapman & Hall, New York, pp. 159–194.
45. Slack, K.V., R.C. Averett, P.E. Greeson et al. 1973. Methods for collection and analysis of aquatic biological and microbiological samples. In *Techniques of Water—Resources Investigations of the United States Geological Survey,* U.S. Department of the Interior, Geological Survey, Washington, D.C., pp. 1–165.

46. Shannon, C.E. 1948. A mathematical theory of communication. *Bell System Technical Journal* 27:379–423.
47. Whittaker, R.H. 1952. A study of summer foliate insect communities in the Great Smoky Mountains. *Ecological Monographs* 22:1–44.
48. Simpson, E.H. 1949. Measurement of diversity. *Nature* 163:688.
49. Washington, H.G. 1984. Diversity, biotic and similarity indices—a review with special relevance to aquatic ecosystems. *Water Research* 18:653–694.
50. Vannote, R.L., G.W. Minshall, and K.W. Cummins et al. 1980. The river continuum concept. *Canadian Journal of Fisheries and Aquatic Science* 37:130–137.
51. Karr, J.R., K.D. Fausch, and P.L. Angermeier et al. 1996. *Assessing Biological Integrity in Running Waters, a Method and Its Rationale,* Special Publication 5, Illinois Natural History Survey.
52. SalmonWeb. *Biological Integrity and the Index of Biological Integrity.* Available at: http://www.salmonweb.org/salmonweb/pubs/biomonitor.html.
53. Karr, J.R. 1981. Assessment of biotic integrity using fish communities. *Fisheries* 6:21–27.
54. Goodman, D. 1975. The theory of diversity–stability relationships in ecology. *Quarterly Review of Biology* 50:237–266.
55. Hurlbert, S.H. 1984. The nonconcept of species diversity: a critique and alternative parameters. *Ecology* 52:577–586.
56. Wilhm, J.L. 1972. Graphic and mathematical analyses of biotic communities in polluted streams. *Annual Review of Entomology* 17:223–252.
57. Hellawell, J.M. 1986. *Biological Indicators of Freshwater Pollution and Environmental Management.* Elsevier, London.
58. Krebs, C.J. 1985. *Ecology—The Experimental Analysis of Distribution and Abundance,* 3rd ed., Harper & Row, New York.
59. Hughes, B.D. 1978. The influence of factors other than pollution on the value of Shannon's Diversity Index for benthic macroinvertebrates in streams. *Water Research* 12:359–364.
60. Barton, D.R. and B.W. Kilgour. 1998. A preliminary evaluation of the behaviour of the BioMAP Water Quality Index. *Canadian Water Resources Journal* 24:139–146.
61. Griffiths, R.W. 2000. Comment on "A Preliminary Evaluation of the Behaviour of the BioMAP Water Quality Index." *Canadian Water Resources Journal* 25:93–101.
62. Hilsenhoff, W.L. 1987. An improved biotic index of organic stream pollution. *Great Lakes Entomology* 20:31–39.
63. Reynoldson, T.B. and D.M. Rosenberg. 1996. Sampling strategies and practical considerations in building reference databases for the prediction of invertebrate community structure. In *Proceedings of the 44th Annual Meeting of the North American Benthological Society, Ninth Annual Technical Information Workshop.* North American Benthological Society, Kalispell, MT.
64. Bailey, R. 1996. Comparing predicted and actual benthic invertebrate communities in test ecosystems. In *Proceedings of the 44th Annual Meeting of the North American Benthological Society, Ninth Annual Technical Information Workshop.* North American Benthological Society, Kalispell, MT.
65. Hughes, R.M. 1995. Defining acceptable biological status by comparing with reference conditions. In *Biological Assessment and Critera: Tools for Water Resource Planning and Decision Making,* S. Davis and T.P. Simons, Eds. Lewis Publishers, Boca Raton, FL, pp. 31–47.

66. Wright, J.F., D. Moss, P.D. Armitage et al. 1984. A preliminary classification of running-water sites in Great Britain based on macro-invertebrate species and the prediction of community type using environmental data. *Freshwater Biology* 14:221–256.
67. Moss, D., M.T. Furse, and J.F. Wright. 1987. The prediction of the macro-invertebrate fauna of unpolluted running-water sites in Great Britain using environmental data. *Freshwater Biology* 17:41–52.
68. Green, R.H. 1979. *Sampling Design and Statistical Methods for Environmental Biologists.* Wiley, New York.
69. Johnson, R.A. and D.W. Wichern. 1992. *Applied Multivariate Statistical Analysis,* 3rd ed., Prentice-Hall, Englewood Cliffs, NJ.
70. Rosenberg, D.M. and V.H. Resh, Eds. 1993. *Freshwater Biomonitoring and Benthic Macroinvertebrates.* Chapman & Hall, New York.
71. Griffiths, R.W. 1993. Statistical applications: matching theory with the reality of benthic surveys. In *Proceedings of the North American Benthological Society, Sixth Annual Technical Information Workshop.* North American Benthological Society, Calgary, Alberta, Canada.
72. U.S. Environmental Protection Agency, Office of Water. 1991. *Technical Support Document for Water Quality Based Toxics Control.* U.S. Environmental Protection Agency, Office of Water, Washington, D.C.
73. British Columbia Ministry of Environment, Environmental Protection Department. *The British Columbia Water Quality Index.* Available at: http://www.env.gov.bc.ca/wat/wq/bcguidelines/indexreport.html.
74. U.S. Environmental Protection Agency, Office of Water. *Total Maximum Daily Load (TMDL) Program.* Available at: http://www.epa.gov/owow/tmdl/.
75. Ontario Ministry of Environment and Energy. 1994. *Water Management Policies, Guidelines, Provincial Water Quality Objectives of the Ministry of Environment and Energy.* Queen's Printer for Ontario, Toronto, Canada.
76. Laws, E.A. 2000. *Aquatic Pollution—An Introductory Text,* 3rd ed., Wiley, New York.
77. Ontario Ministry of Environment and Energy, Customer Services Unit, Laboratory Services Branch. 1993. *A Guide to the Collection and Submission of Samples for Laboratory Analysis,* 7th ed., Ontario Ministry of the Environment and Energy, Toronto, Canada.
78. Canadian Council of Ministers of the Environment. 1994. *Subsurface Assessment Handbook for Contaminated Sites,* Report CCME EPC-NCSRP-48E. National Printers, Ottawa, Canada.
79. Chatfield, C. 1996. *The Analysis of Time Series, An Introduction,* 5th ed. Chapman & Hall, London.
80. Hunsacker, C.T. and D.A. Levine. 1995. Hierarchical approaches to the study of water quality in rivers. *Bioscience* 45:193–203.
81. Intergovernmental Task Force on Monitoring Water Quality (ITFM). 1992. *Ambient Water Quality Monitoring in the United States. First Year Review, Evaluation, and Recommendations.* ITFM, Interagency Advisory Committee on Water Data, Water Information Coordination Program, U.S. Geological Survey, Washington, D.C.
82. Intergovernmental Task Force on Monitoring Water Quality. 1995. *The Strategy for Improving Water-Quality Monitoring in the United States: Final Report of the Intergovernmental Task Force on Monitoring Water Quality.* U.S. Geological Survey, Reston, VA.

83. Intergovernmental Task Force on Monitoring Water Quality. 1995. *The Strategy for Improving Water-Quality Monitoring in the United States: Final Report of the Intergovernmental Task Force on Monitoring Water Quality, Technical Appendices.* U.S. Geological Survey, Reston, VA.
84. U.S. Environmental Protection Agency. 1996. *Nonpoint Source Monitoring and Evaluation Guide.* U.S. Environmental Protection Agency, Office of Water, Washington, D.C.
85. Stanfield, L., M. Jones, and M. Stoneman. 1997. *Stream Assessment Protocol for Southern Ontario.* Ontario Ministry of Natural Resources, Picton, Canada.
86. U.S. Department of the Interior and U.S. Geological Survey. *Habitat Suitability Index (HSI) Models for Selected Fish and Wildlife Species.* Available at: http://www.mesc.usgs.gov/hsi/hsi_models_available.html.
87. South Dakota State University and South Dakota Cooperative Fish and Wildlife Research Unit. *Insert Flow Incremental Methodology, Habitat Suitability Indices, and Physical Habitat Simulation Modeling (IFIM, HSI, and PHabSim).* Available at: http://wfs.sdstate.edu/wl719/lectures/20ifim/tsld001.htm.
88. Covich, A.P., W.H. Clements, and K.D. Fausch et al. (Colorado Water Resources Research Institute). *Measurable Parameters of Ecological Integrity.* Available at: http://cwrri.colostate.edu/pubs/balance/no.3/bal3.html.
89. Rasmussen, J.L. (American Fisheries Society). *American Fisheries Society Resource Policy Handbook,* 1st ed. Available at: http://www.fisheries.org/resource/resource.htm.
90. Harrelson, C.C., C.L. Rawlins, and J.P. Potyondy. 1994. *Stream Channel Reference Sites: An Illustrated Guide to Field Technique,* General Technical Report RM-245. USDA Forest Service, Rocky Mountain Forest and Range Experiment Station, Fort Collins, CO.
91. Rosgen, D. 1996. *Applied River Morphology.* Wildland Hydrology, Pagosa Springs, CO.
92. North Carolina State University Stream Restoration Institute. *Conference Proceedings, Stream Restoration & Protection in North Carolina: Building on Success.* Available at: http://www.bae.ncsu.edu/programs/extension/wqg/sri/proceedings.htm.
93. Schuett-Hames, D. and A. Pleus (Northwest Indian Fisheries Commission). *Literature Review & Monitoring Recommendations for Salmonid Spawning Habitat Availability.* Available at: http://www.nwifc.wa.gov/tfw/reports/report3.htm.
94. Barbour, M.T. and J.B. Stribling. 1991. Use of habitat assessment in evaluating the biological integrity of stream communities. In *Biological Criteria: Research and Regulation, Proceedings of a Symposium,* 12–13 December 1990, Arlington, Virginia, G. Gibson, Ed., EPA-404-5-91-005. U.S. Environmental Protection Agency, Office of Water, Washington, D.C., pp. 25–38.
95. Barbour, M.T. and J.B. Stribling. 1994. A technique for assessing stream habitat structure. In *Conference Proceedings, Riparian Ecosystems in the Humid U.S.: Functions, Values and Management.* National Association of Conservation Districts, Washington, D.C., pp. 156–178.
96. Beschta, R.L. and W.S. Platts. 1996. Morphological features of small streams—significance and function. *Water Resources Bulletin* 22:369–379.
97. Gordon, N.D., T.A. McMahon, and B.L. Finlayson. 1992. *Stream Hydrology: An Introduction for Ecologists.* Wiley, West Sussex, U.K.

98. Hannaford, M.J. and V.H. Resh. 1995. Variability in macroinvertebrate rapid-bioassessment surveys and habitat assessments in a northern California stream. *Journal of the North American Benthological Society* 14:430–439.
99. Hannaford, M.J., M.T. Barbour, and V.H. Resh. 1997. Training reduces observer variability in visual-based assessments of stream habitat. *Journal of the North American Benthological Society* 16:853–860.
100. Hawkins, C.P., J.L. Kershner, P.A. Bisson et al. 1993. A hierarchical approach to classifying stream habitat features. *Fisheries* 18:3–12.
101. Kaufmann, P.R. 1993. Physical habitat. In *Stream Indicator and Design Workshop Pages*, R.M. Hughes, Ed., EPA/600/R-93/138. U.S. Environmental Protection Agency, Corvallis, OR, pp. 59–69.
102. Kaufmann, P.R. and E.G. Robison. 1997. Physical habitat assessment. In *Environmental Monitoring and Assessment Program, 1997 Pilot Field Operations Manual for Streams,* D.J. Klemm and J.M. Lazorchak, Eds., EPA/620/R-94/004. U.S. Environmental Protection Agency, Office of Research and Development, Environmental Monitoring Systems Laboratory, Cincinnati, OH, pp. 6-1-6-38.
103. Meador, M.R., C.R. Hupp, T.F. Cuffney et al. 1993. *Methods for Characterizing Stream Habitat as Part of the National Water-Quality Assessment Program,* Open-File Report USGS/OFR 93–408. U.S. Geological Survey, Raleigh, NC.
104. Oswood, M.E. and W.E. Barber. 1982. Assessment of fish habitat in streams: goals, constraints, and a new technique. *Fisheries* 7:8–11.
105. Rankin, E.T. 1991. The use of the Qualitative Habitat Evaluation Index for use attainability studies in streams and rivers in Ohio. In *Biological Criteria: Research and Regulation,* G. Gibson, Ed., EPA 440/5–91–005. U.S. Environmental Protection Agency, Office of Water, Washington, D.C.
106. Rankin, E.T. 1995. Habitat indices in water resource quality assessments. In *Biological Assessment and Criteria: Tools for Water Resource Planning and Decision Making,* W.S. Davis and T.P Simon, Eds. Lewis Publishers, Boca Raton, FL, pp. 181–208.
107. Raven, P.J., N.T.H. Holmes, F.H. Dawson et al. 1998. *River Habitat Quality: The Physical Character of Rivers and Streams in the UK and Isle of Man.* Environment Agency, Bristol, U.K.
108. Rosgen, D.L. 1985. A stream classification system. In *Proceedings of the First North American Riparian Conference, Riparian Ecosystems and Their Management: Reconciling Conflicting Uses,* General Technical Report RM-120. U.S. Department of Agriculture Forest Service, Tucson, AZ, pp. 91–95.
109. Simonson, T.D., J. Lyons, and P.D. Kanehl. 1994. Quantifying fish habitat in streams: transect spacing, sample size, and a proposed framework. *North American Journal of Fisheries Management* 14:607–615.
110. Stribling, J.B., B.K. Jessup, and J. Gerritsen. 1999. *Development of Biological and Habitat Criteria for Wyoming Streams and Their Use in the TMDL Process.* Tetra Tech, Owings Mills, MD.
111. Wesche, T.A., C.M. Goertler, and C.B. Frye. 1985. Importance and evaluation of instream and riparian cover in smaller trout streams. In *Proceedings of the First North American Riparian Conference Riparian Ecosystems and Their Management: Reconciling Conflicting Uses,* General Technical Report TM-120. U.S. Department of Agriculture Forest Service, Tucson, AZ, pp. 325–328.
112. Pierce, R.C., D.M. Whittle, and J.B. Bramwell, Eds. 1998. *Chemical Contaminants in Canadian Aquatic Ecosystems.* Minister of Public Works and Government Services Canada, Ottawa, Ontario, Canada.

113. Cable News Network. *Group Calls for Worldwide DDT Ban.* Available at: http://www.cnn.com/tech/science/9901/29/ddt.enn/.
114. Government of Canada. 1991. *Toxic Chemicals in the Great Lakes and Associated Effects.* Government of Canada, Ottawa, Ontario, Canada.
115. Environment Canada. 1994. *Guidance Document on Collection and Preparation of Sediments for Physicochemical Characterization and Biological Testing,* Report EPS 1/RM/29. Environment Canada, Technology Development Directorate, Ottawa, Ontario.
116. U.S. Environmental Protection Agency. 1985. *Technical Support Document for Water Quality–Based Toxics Control.* U.S. Environmental Protection Agency, Washington, D.C.
117. Lee, G.F. and A. Jones-Lee. *Guidance for Conducting Water Quality Studies for Developing Control Programs for Toxic Contaminants in Wastewaters and Stormwater Runoff.* Available at: http://home.pacbell.net/gfredlee/stdy-app.htm.
118. Mudroch, A. and S.D. MacKnight. 1991. *Handbook of Techniques for Aquatic Sediments Sampling.* CRC Press, Boca Raton, FL.
119. North Carolina State University. *Multiple Regression.* Available at: http://www2.chass.ncsu.edu/garson/pa765/regress.htm.
120. San José State University. *Continuous Outcome, Multiple Predictors (Multiple Regression).* Available at: http://www.sjsu.edu/faculty/gerstman/epiinfo/cont-mult.htm.
121. Warner, J., D. Steppan, and B. Yeater. *Essential Regression and Experimental Design.* Available at: http://www.geocities.com/siliconvalley/network/1900/.
122. U.S. Environmental Protection Agency. *The Volunteer Monitor's Guide To Quality Assurance Project Plans.* Available at: http://www.epa.gov/owow/monitoring/volunteer/qappexec.html.
123. Gauthier, L.G. 1998. Introduction to GIS. In *Proceedings of the North American Benthological Society Technical Workshop.* North American Benthological Society, Charlottetown.
124. Gauthier, L.G. 1998. Data sources/data quality. In *Proceedings of the North American Benthological Society Technical Workshop.* North American Benthological Society, Charlottetown, Prince Edward Island, Canada.
125. U.S. Environmental Protection Agency, Office of Water. *Designing an Information Management System for Watersheds.* Available at: http://www.epa.gov/owow/watershed/wacademy/its05/index.html.
126. University of Arkansas. *Japan GIS/Mapping Sciences Resource Guide,* 3rd ed. Available at: http://www.cast.uark.edu/jpgis/jpsoft.html.
127. Adams Business Media. *GeoWorld GIS Software.* Available at: http://www.geoplace.com/gw/1999/0799/799gis.asp.
128. U.S. Environmental Protection Agency. *Mapping Water Quality Information with Reach File 3—A Case Study.* Available at: http://www.epa.gov/ceisweb1/ceishome/atlas/hydrography/mapping wqinfo.html.
129. New York State Archives. *GIS Glossary.* Available at: http://www.archives.nysed.gov/pubs/gis/glossary.htm.
130. University of Northern British Columbia. *Geog 205 & Geog 300 Glossary (Glossary of GIS Terms).* Available at: http://www.gis.unbc.ca/webpages/start/resources/glossary.html.

131. Korte, G.B. 1997. *The GIS Book: Understanding the Value and Implementation of Geographic Information Systems,* 4th ed. Onword Press, Albany, NY.
132. James, W. *List of Programs and Models (GIS Programs).* Available at: http://www.eos.uoguelph.ca/webfiles/james/wjrefprograms.html.
133. Sinclair, W. and R.W. Pressinger. *Many Illnesses Suspected for People Living in Chlordane Pesticide Treated Homes.* Available at: http://www.main.nc.us/pace/chlordane.html.
134. U.S. Environmental Protection Agency, History Office. *DDT Ban Takes Effect.* Available at: http://www.epa.gov/history/topics/ddt/01.htm.
135. Ritter, L., K.R. Solomon, J. Forget et al. *An Assessment Report On: DDT, Aldrin, Dieldrin, Endrin, Chlordane, Heptachlor, Hexachlorobenzene, Mirex, Toxaphene, Polychlorinated Biphenyls, Dioxins and Furans.* Available at: http://www.chem.unep.ch/pops/indxhtms/asses0.html#toc.
136. Environmental Defense Newsletter. *Court Upholds Aldrin–Dieldrin Ban.* Available at: http://www.edf.org/pubs/edf-letter/1975/may/b_aldrin.html.
137. U.S. Environmental Protection Agency and Environment Canada. *The Great Lakes Binational Toxics Strategy. Canada–United States Strategy for the Virtual Elimination of Persistent Toxic Substances in the Great Lakes.* Available at: http://www.epa.gov/glnpo/p2/bns.html.
138. U.S. Department of Health and Human Services, Public Health Service, Agency for Toxic Substances and Disease Registry. *Hexachlorocyclohexane.* Available at: http://www.atsdr.cdc.gov/tfacts43.html.
139. Wild, G. and J. De Koning (ERS Department, Trent University). *Contaminants: Information on the Chemical Structure and Properties of Organochlorine Pesticides and PCBs.* Available at: http://www.whalenet.org/bwcontaminants/contaminants.html.
140. University of Kassel. *Agricultural Non-Point Source Pollution Model, General Model Information.* Available at: http://www.wiz.uni-kassel.de/model_db/mdb/agnps.html.
141. Young, R., C.A. Onstad, D.D. Bosch et al. 1987. *AGNPS: Agricultural Non-Point Source Pollution Model: A Watershed Analysis Tool,* Conservation Research Report 35. U.S. Department of Agriculture, Agricultural Research Service, Washington, D.C.
142. Young R.A., C.A. Onstad, D.D. Bosch et al. 1989. AGNPS: A non-point source pollution model for evaluating agricultural watersheds. *Journal of Soil and Water Conservation* 44: 168–173.
143. Battin, A., R. Kinerson, and M. Lahlou. *EPA's Better Assessment Science Integrating Point and Nonpoint Sources (BASINS), A Powerful Tool for Managing Watersheds.* Available at: http://www.crwr.utexas.edu/gis/gishydro99/epabasins/battin/p447.htm.
144. U.S. Environmental Protection Agency, Office of Science and Technology. *BASINS—Better Assessment Science Integrating Point and Nonpoint Sources.* Available at: http://www.epa.gov/ost/basins/.
145. U.S. Army Corps of Engineers. *Object-Oriented GAWSER Model.* Available at: http://www.crrel.usace.army.mil/rsgisc/gwsr.htm.
146. Grand River Conservation Authority, Geographical Information Systems. *GIS Report.* Available at: http://www.grandriver.on.ca/gis/grca_gn4.htm.
147. U.S. Environmental Protection Agency. *Center For Exposure Assessment Modeling—SWMM.* Available at: http://www.epa.gov/ceampubl/swmm.htm.

148. Oregon State University, Department of Civil, Construction, and Environmental Engineering. *EPA Storm Water Management Model (SWMM), Versions 4.31 and 4.4.* Available at: http://www.ccee.orst.edu/swmm/.
149. Huber, W.C., J.P. Heaney, and B.A. Cunningham. 1985. *Storm Water Management Model (SWMM) Bibliography,* EPA/600/3–85/077, NTIS PB86–136041. U.S. Environmental Protection Agency, Athens, GA.
150. Greenland International Consulting Inc. *ISWMS Homepage.* Available at: http://www.grnland.com/projects/iswms/iswms.htm.
151. Ministry of Transportation Ontario. *Selecting Computational Methods.* Available at: http://www.mto.gov.on.ca/english/engineering/drainage/section10.htm.
152. Ontario Ministry of Environment and Energy. 1993. *Watershed Planning.* Queen's Printer for Ontario, Ontario, Canada.
153. Environmental Commissioner of Ontario. 2000. *Changing Perspectives.* Queen's Printer for Ontario, Toronto, Ontario.
154. City of Waterloo. 1995. *Strategic Plan West Side Watershed Implementation Discussion Paper,* Report P& PW/EG4254, City of Waterloo, Ontario, Canada.
155. Grand River Conservation Authority. 1993. *Laurel Creek Watershed Study,* Grand River Conservation Authority, Cambridge, Ontario, Canada.
156. Region of Waterloo. 1996. *Regional Official Policies Plan,* City of Waterloo, Ontario, Canada.
157. City of Waterloo. 1994. *Municipal Official Plan,* City of Waterloo, Ontario, Canada.
158. City of Waterloo. 1996. *City of Waterloo District Implementation Plan,* Columbia Hills, Ontario, Canada, City of Waterloo, Ontario, Canada.
159. United Nations Centre for Human Settlements (Habitat). *Best Practices and Local Leadership Program.* Available at: http//www.sustainabledevelopment.org/blp.
160. Best Practices for Human Settlements. *Best Practices Database.* Available at: http//www.bestpractices.org.
161. Dubai International Award for Best Practices to Improve the Living Environment. *Best Practices Database.* Available at: http//www.dubai-award.dm.gov.ae/.
162. United Nations Centre for Human Settlements. *Best Practices & Local Leadership Program.* Available at: http//www.habitat.unchs.org/home.htm.
163. Lake Simcoe Region Conservation Authority. 1997. *Uxbridge Brook Watershed Plan.* Lake Simcoe Region Conservation Authority, Newmarket, Ontario, Canada.
164. Heathcote, I.W. 1998. *Integrated Watershed Management: Principles and Practices.* Wiley, New York.
165. Walters, M.J. and A. Westwood. 1989. *Clean Up Rural Beaches Plan—Pefferlaw Brook Drainage Basin.* Lake Simcoe Region Conservation Authority, Newmarket, Ontario, Canada.
166. Senes Consultants Ltd. 1994. *Surface Water Quality Assessment Report for Uxbridge Brook.* Regional Municipality of Durham and Township of Uxbridge, Ontario, Canada.
167. Stockwell, S. and M. Jones. 1994. *Single-Pass Method Stream Assessment Protocol.* Ontario Ministry of Natural Resources, Picton, Canada.
168. Stoneman, M. and M. Jones. 1993. *Methods for Assessing the Status of Coldwater Streams.* Draft Report. Ontario Ministry of Natural Resources, Picton, Canada.
169. BEAK Consultants. 1995. *Development and Implementation of a Phosphorus Loading Watershed Management Model for Lake Simcoe,* LSEMS Implementation Technical Report A.3. Lake Simcoe Region Conservation Authority, Newmarket, Ontario, Canada.

170. Weatherbe, D.G. and J. Li. 2000. *Uxbridge Urban Area Stormwater Management Study*. Township of Uxbridge, Uxbridge, Ontario, Canada.

171. Palmer, R.M., C. Jones, and M. Walters. 1998. Environmental Monitoring Initiatives to Sustain Growth in Ontario, Canada. *Journal of Water Science and Technology* 38:113–122; Palmer, R.M., C. Jones, and M. Walters. 1998. Environmental Monitoring Initiatives to Sustain Growth in Ontario, Canada. In *Proceedings of the International Association on Water Quality 19th Biennial Conference*, International Association of Water Quality, Vancouver, British Columbia, Canada.

172. Lei, J., Li, J., and W. Schilling. 1999. *Stepwise hypothesis test model calibration procedure of urban runoff design model and an alternative of Monte-Carlo simulation*. Proceedings of the 8th International Conference on Urban Storm Drainage, Sydney, Australia, 30 August–3 September, ISBN 0-85825-718-1, pp. 973-981.

173. Gupta, H.V., S. Sorooshian, and P.O. Yapo. 1998. Toward improved calibration of hydrological models: multiple and noncommensurable measures of information. *Water Resources Research* 34:751–763.

174. Beven, K.J. 1993. Prophecy, reality and uncertainty in distributed hydrological modeling. *Advances in Water Resources* 16:41–51.

175. Greenland International Consulting Inc. 2001. *A-E-M-O-T Groundwater Management Study—Final Report*. Greenland International Consulting Inc., Toronto, Ontario, Canada.

176. Jobin, D.I. 1998. *Water Management Plan for the Mississippi River Watershed—Phase 1: Hydrologic Data Optimization Using RADARSAT Project Report*.

177. Peat, G. and M.J. Walters. 1994. *Lake Simcoe Tributary Monitoring Data Report*, LSEMS Report Imp.A.2. Lake Simcoe Region Conservation Authority, Newmarket, Ontario, Canada.

178. Scott, L.D., J.G Winter, M.N. Futter et al. 2001. *Annual Water Balance and Phosphorus Loading for Lake Simcoe, 1990–1998*, LSEMS Report Imp.A.4. Lake Simcoe Region Conservation Authority, Newmarket, Ontario, Canada.

179. McMurtry, M.J., C.C. Willox, and T.C. Smith. 1997. An overview of fisheries management for Lake Simcoe. *Journal of Lake and Reservoir Management* 13:199–213.

180. MacLean, J.A., Evans, D.O., Martin, N.V. et al. 1981. Survival, growth, spawning distributions and movements of introduced native lake trout (*Salvlinus namaycush*) in two inland Ontario lakes. *Canadian Journal of Fisheries and Aquatic Science* 38:1685–1700.

181. Evans, D.O., K.H. Nicholls, Y.C. Allen et al. 1996. Historical land use, phosphorus loading, and loss of fish habitat in Lake Simcoe, Canada. *Canadian Journal of Fisheries and Aquatic Science* 53 (Suppl. 1):194–218.

182. Snodgrass, W.J. and J. Holubshen. 1993. *Hypolimnetic Oxygen Dynamics in Lake Simcoe, Part 3: Model Confirmation and Prediction of the Effects of Management*, LSEMS Report Imp.B.16. Lake Simcoe Region Conservation Authority, Newmarket, Ontario, Canada.

183. BEAK Consultants. 1994. *Development and Implementation of a Phosphorus Loading Watershed Management Model for Lake Simcoe. Draft Report*. Lake Simcoe Region Conservation Authority, Newmarket, Ontario, Canada.

184. Walker, R.R. 1997. Strategies to rehabilitate Lake Simcoe: phosphorus loads, remedial measures and control options. *Journal of Lake and Reservoir Management* 13(4):214–225.

185. Nicholls, K.H. 1995. *A Limnological Basis for a Lake Simcoe Phosphorus Loading Objective,* LSEMS Report Imp.B.17. Lake Simcoe Region Conservation Authority, Newmarket, Ontario, Canada.
186. Dillon, P.J. and F.H. Rigler. 1974. The phosphorus–chlorophyll relationship in lakes. *Limnology and Oceanography* 19:767–773.
187. Dillon, P.J. and F.H. Rigler. 1975. A simple method for development of predicting the capacity of a lake for development based on lake trophic status. *Journal of Fishery Resource Board of Canada* 32:1519–1531.
188. Vollenweider, P.A. 1975. Input–output models with special reference to the phosphorus loading concept in limnology. *Swiss Journal of Hydrology* 37:53–84.
189. Vollenweider, P.A. 1976. Advances in defining critical loading levels for phosphorus in lake eutrophication. *Memorie. Istituto Italiano di Idrobiologia* 33:53–83.
190. Vollenweider P.A. and J.J. Kerkes. 1980. The Loading Concept as Basis for Controlling Eutrophication: Philosophy and Preliminary Results of the OECD Programme on Eutrophication. *Progress in Water Technology* 12. IAWPP/Pergamon Press, New York, pp. 5–38.
191. Janus L.L. and P.A. Vollenwider. 1981. *Summary Report: The OECD Cooperative Programme on Eutrophication, Canadian Contribution,* Scientific Series No. 131. National Water Research Institute, Inland Waters Directorate, Burlington, Ontario, Canada.
192. Smith V.H. and J. Shapiro. 1981. Chlorophyll–phosphorus relations in individual lakes. Their importance to lake restoration strategies. *Environmental Science and Technology* 15:444–451.
193. Johanson, R.C., J.C. Imhof, J.L. Kittle, Jr. et al. 1984. *Hydrologic Simulation Program—Fortran (HSPF) User's Manual for Release 8.0,* Report EPA-600/3-84-066. Environmental Research Laboratory, Athens, GA.
194. Lake Simcoe Environmental Management Strategy Steering Committee. 1985. *Final Report and Recommendations of the Steering Committee.* Lake Simcoe Region Conservation Authority, Newmarket, Ontario, Canada.
195. U.S. Environmental Protection Agency. 1996. *Draft Framework for Watershed Based Trading.* Office of Water, Washington, D.C.
196. Draper, D.W. and M. Fortin. 2000. *Total Phosphorus Management in Lake Simcoe.* Lake Simcoe Region Conservation Authority, Newmarket, Ontario, Canada.
197. Cluis, D., C. Langlois, R. Van Coillie, et al. 1989. Development of a software package for trend detection in temporal series: application to water and industrial effluent quality data for the St. Lawrence River. *Journal of Environmental Monitoring and Assessment,* Available at: http://Kluweronline.com.
198. Nicholls, K.H. 1998. *Lake Simcoe Water Quality Update with Emphasis on Phosphorus Trends,* LSEMS Report Imp.B.17. Lake Simcoe Region Conservation Authority, Newmarket, Ontario, Canada.
199. Evans, D.O. and P. Waring. 1987. Changes in multispecies winter angling fishery of Lake Simcoe, Ontario, 1961–83: invasion by Rainbow Smelt *Osmerus mordax* and the roles of intra- and inter-specific interactions. *Canadian Journal of Fisheries and Aquatic Science* 44(Suppl. 2):182–197.
200. BEAK Consultants. 1995. *Development and Implementation of a Phosphorus Loading Watershed Management Model for Lake Simcoe,* LSEMS Implementation Technical Report A.3, Lake Simcoe Region Conservation Authority, Newmarket, Ontario, Canada.

# Index

## A



## B

# D

## E



## F

## G



## H



## I

## J



## K



## L

# M

# N



# O



# P

## V



## W



## Y

# Z